The Silkewormes and their Flies

medieval & renaissance texts & studies

VOLUME 61

RENAISSANCE ENGLISH TEXT SOCIETY
SIXTH SERIES
VOL. XIII (1988)

The Silkewormes and their Flies

Thomas Moffet

(1599)

A Facsimile
Edited with Introduction and Commentary
by

VICTOR HOULISTON

mediEval & REnaissance texts & studies

in conjunction with
Renaissance English Text Society
Binghamton, New York
1989

© Copyright 1989
Center for Medieval and Early Renaissance Studies
State University of New York at Binghamton

Library of Congress Cataloging-in-Publication Data

Moffet, Thomas, 1553-1604.
    The silkewormes and their flies.

    (Medieval & Renaissance texts & studies ; vol. 56)
    Includes index.
    1. Silkworms—Poetry.   2. Sericulture—Poetry.
I. Houliston, Victor, 1954-   . II. Title.
III. Series: Medieval & Renaissance texts & studies ; v. 56.

PR2315.M65S5   1989   821'.3   88-8553
ISBN 0-86698-040-7 (alk. paper)

This book is made to last.
It is set in Bembo, smythe-sewn,
and printed on acid-free paper
to library specifications.

Printed in the United States of America

To my wife, Margaret

Acknowledgements

I gratefully acknowledge my debts to Dr. Jonquil Bevan, Professor Albert Braunmuller, Dr. Michael Brennan, Miss Katherine Duncan-Jones, Professor Mario A. Di Cesare, Professor John Gouws, Professor Cecil Grayson, Professor Arthur F. Kinney, Mr. Michael Piret, Dr. John Pitcher, Dr. Robin Robbins, and Dr. Charles Webster, for assistance and advice in the preparation of this edition. All errors are, of course, my own responsibility.

The research for this edition was funded by the trustees of the Max and Lillie Sonnenberg Scholarship, the Scarbrow Fund, the Meyerstein Fund and the Underhill Fund, to whom my thanks are due.

Contents

Acknowledgements	VI
Abbreviations and Conventions	VIII
General Introduction	XI
Date	XVI
Authorship	XVIII
Text	XIX
Notes to the Introduction	XXV
The Silkewormes and their Flies	1
Apparatus Criticus	79
Commentary	83
Glossary	95
Index to the Commentary	101

Abbreviations and Conventions

In all citations, the place of publication is London unless otherwise noted. References to classical works normally use the system associated with each work in the Loeb Classical Library. For early printed books I have where possible given both a chapter and a page reference as the edition used may not be readily available to the reader. Where classical and contemporary Renaissance works other than those cited by Moffet are referred to in my commentary, the intention is to indicate the currency of a notion, rather than to identify sources.

References to Sidney's poems follow the system of abbreviation formulated by W. A. Ringler in his edition of 1962.

C.S.P.	*Calendar of State Papers* (Public Record Office)
du Bartas	Sallust du Bartas, *Works*, ed. U. T. Holmes et al. (Chapel Hill, 1935–40). Line references are to the French, unless the Sylvester translation is specified.
(trans. Sylvester)	*The Divine Weeks and Works*, trans. Joshua Sylvester (1605), ed. Susan Snyder (Oxford English Texts, 2 vols, 1979).
H.M.C.	*Historical Manuscripts Commission*
Healths Improvement	Moffet, Thomas, *Healths Improvement* (1655).
Scoular	Scoular, Kitty, *Natural Magic* (1965).
S.T.C.	*A Short-Title Catalogue of Books 1475–1640*, ed. A. W. Pollard and G. R. Redgrave (1926).
Tilley	Tilley, M. P., *Dictionary of Proverbs in England in the Sixteenth and Seventeenth Centuries* (Ann Arbor, 1950).
Vida	Vida, Marcus Hieronymus, *De bombyce* (1527; line numbers refer to the Latin text in *The Silkworm*, trans. S. Pullein [Dublin, 1750]).

The Silkewormes and their Flies

General Introduction

The Silkewormes and their Flies was the first published attempt to promote the planting of mulberry trees and the rearing of silkworms in England and the only full-length poem on the subject. But it reflected widespread contemporary interest. The demand for silk stockings and other products increased considerably in the late Elizabethan period and it seemed wasteful to rely on imports from Italy and Spain.[1] James I accordingly supported a mulberry tree-planting campaign, sponsoring pamphlets by Nicholas Geffe, *The perfect use of silkwormes, and their benefit* (1607; actually a translation of a work by Olivier de Serres; there are dedicatory verses by Michael Drayton), and William Stallenge, *Instructions for the increasing of mulberie trees, and the breeding of silke-wormes* (1609). Henry Peacham commended Stallenge as the "first Author of making Silke in our land,"[2] and Charles Butler, writing a bee-keeping treatise in 1609, acknowledged that "the delicat Silk-worme ... is now setting foote in this land."[3] But when Michael Drayton's patron Walter Aston was granted the office of Keeper of the royal Mulberry Garden in 1629, he did not consider it an adequate recompense for his efforts as ambassador to Spain.[4]

The Silkewormes has a strong claim to the title of the first Vergilian georgic in English. Its closest rival is probably John Dennys's *The Secrets of Angling* (before 1609, the year of the author's death), in three books. This poem, also in *ottava rima*, bears many similarities to *The Silkewormes* in versification and diction, although it is in places much more lyrical and reflective. Whatever may be said about the presence of georgic elements in Tusser, Gavin Douglas, or Spenser, Moffet is unique in that he advertises his poem as georgic. His title page gives the author's name as "T. M. *a Countrie Farmar and an ap-*prentice in physicke" and states the intention, "*For the great benefit and enriching of England.*" He models the poem on Vergil and neo-Latin georgic, and explicitly detaches himself from the fictional mode of poetry epitomized in the name of the very man, Philip Sidney, whose muse is invoked in the opening lines. Of the various Ovidian-style accounts of the origin of silk, he remarks

> These be the tales that Poetizers sing ...
> Sweete, I confesse, and drawn from Helique spring,
> Full of delighting change, and learning greate.
> Yet, yet, my Muse dreames of another thing,
> And listeth not of fictions to entreate. (p. 20)

The erudition displayed in the poem may seem more superficial than real, especially since some of the marginal references are inaccurate and others misleading (for example, those to Bonfinis and Pausanias, p. 68), but Moffet was indeed widely read. As a natural historian, he was steeped in Pliny and in Conrad Gesner's *Historia animalium* (4 vols., Zurich, 1555–58). Although he was a Paracelsian apologist, his medical reading was well balanced, covering the whole range of classical, Arabic, medieval and Renaissance authorities, and he took a special interest in herbals, particularly the *Niewe herball* of Rembert Dodoens (trans. Henry Lyte, 1578). His Cambridge studies account for his extensive classical knowledge. In addition, it is clear from his anecdotal dietary treatise *Healths Improvement* (posthumously published in 1655) that he was an enthusiastic reader of history and biography, evidenced in *The Silkewormes* by marginal references to Polydore Vergil's *De rerum inventoribus libri octo,* Antonius de Bonfinis's *Rerum Hungaricum decades quatuor cum dimidia,* and the encyclopaedic collections of L. Coelius Richerius and M. Coccius Sabellicus.

The general influence of Vergil's fourth *Georgic* can be detected not only in the diction of *The Silkewormes* but also in the sympathetic characterization of the insects, the thematic structure, and the multiple digressions. But Moffet's poem is much more heavily dependent, in every respect, on the *De bombyce* of Marcus Hieronymus Vida (Rome, 1527).[5] An earlier poem on the subject, *Bombyx* by

Ludovicus Lazarellus, was published in Rome about 1495, and reprinted in a collection of *Pictorii epigrammata* in Basle in 1518. Just over 250 lines long, it deals with the care of the silkworm during its life cycle and is distinguished by its explicit religious symbolism. Vida (1485–1566) deals with much the same material as Lazarellus, but his poem, in two books, is almost four times the length, includes mythological narratives and a section on silk manufacture, and eschews religious interpretation. Instead, the atmosphere of his poem is suffused with the presence of the *Seriades*, the silk-nymphs who serve Isabella d'Este, the Marchioness of Mantua. It is to her that the poem is dedicated, and she is addressed directly at the beginning of both books.

Vida's division conforms to a didactic scheme: Book I on the life history of the worm, its feeding and accommodation; Book II on human intervention in curing diseases, handling the cocoons and manufacturing the cloth. Moffet follows a more "epideictic" rhetorical sequence, writing as it were a formal *laus bombycis:* the antiquity and excellence of the worm feature in Book I, its usefulness and benefit in Book II. Accordingly, what material he lifts from Vida is concentrated in Book II of *The Silkewormes*. Even so, he follows a sequence that is slightly more thematic than practical. He inserts learned discussions on alchemy, simplicity of diet (as debated in Plutarch), and more contemporary issues of dress and rank. The whole is rounded off with a "deliberative" oration to persuade husbandmen to plant mulberry trees and engage in sericulture. Of specific debts to Vida, the most interesting from a literary point of view is the legend of Saturn and Phyllira (pp. 5–8), a story which in Moffet's version becomes a vehicle for a most unflattering portrayal of Venus.

Vida's influence partly explains the pastoral frame in which *The Silkewormes* is set, with the Countess of Pembroke and her attendants being addressed by pastoral names (p. 40) and Amyntas and Phillis being reassured that the silkworm will not threaten the shepherds' livelihood (p. 75). But if Moffet does not restrict himself to the conventions of the classical and neo-classical georgic, it is also because *The Silkewormes* is a poem of natural history subject to a variety of non-georgic influences, particularly du Bartas. Spenser, Sidney and Drayton were all admirers of du Bartas, and Sidney is reported to have translated *La premiere sepmaine*.[6] Since Moffet himself mentions this translation in his biography of Sidney, *Nobilis* (p. 74), it is possible that his own interest was kindled by his association with the Pembroke circle, which maintained its enthusiasm for du Bartas well after Philip Sidney's death.[7] In *The Silkewormes*, Moffet enthusiastically develops du Bartas's notion that God's creation is most to be admired in the smallest of creatures; hence the four copious pages (34–37) extolling "the vale" (little things), borrowing in detail a description of Regiomontanus and his mechanical flea. Other allusions to du Bartas in the poem (e.g. pp. 49.1–8; 52.17–18) serve to associate the silkworms (especially their chastity and fidelity) and the mulberry tree with prelapsarian innocence and the marvels of divine providence.

A survey of insect poetry in English in the sixteenth century highlights the objectivity of Moffet's interest and reveals the potential charm of the miniature world of the insect. John Heywood's *The spider and the flie, a parable* (1556) is an ambitious attempt to cast the political history of the mid-century in an allegorical frame as a lasting poetic commentary, but has little to do with natural history. Spenser's two insect poems, *Virgils Gnat* and *Muiopotmos*, unmistakably reflect, like Vida's *De bombyce*, the ambitions of a literary career: the gnat, the butterfly and the spider appear as striking poetic images rather than as animals, their presence dominated by moral and physiological conceits. Tailboys Dymoke's curious poem *Caltha Poetarum* ("Marigold of the Poets," 1599), which relates in 187 seven-line stanzas the eventful romance between a bumblebee and a marigold, testifies to the appeal of an erotically-charged pastoral world of "insects, floral ladies and a rather unbridled set of deities."[8] When Moffet introduces the Ovidian mood into *The Silkewormes* in the episode of Pyramus and Thisbe (pp. 9–19), therefore, he is not simply exploiting all the material he can muster in connection, however tenuous, with silkworms. He can expect an audience familiarized with an easy commerce between the insect and human worlds, and the afterlife.

The author of an insect poem naturally anticipates its classification with trivia, even though he may employ various devices to insist on its serious claims. In versifying the life of the silkworm, Moffet accordingly draws extensively on the venerable rhetorical tradition of the paradoxical encomium. Like Lucian's *The Fly (Muscae laudatio)*, with which Moffet was undoubtedly familiar,[9] *The Silkewormes* has recognizable affinities with the funeral oration, paying tribute to the "gallant" worms that "die" in their cocoons (pp. 61–62) and to the moths that die after copulation:

> Go worthy soules (so witty *Greeks* you name)
> Possesse for aye the faire *Elisian* greene:
>
> Weepe not faire *Mira* for this funeral. (p. 40)

Moffet's own scientific and medical background contributed another encomiastic element: the rivalry between sericulture and alchemy. In particular, he plays the silkworm against the alchemists' "quintessence," comparing its eggs to the philosophers' egg, and its silk to the golden fleece and the golden apples of the Hesperides, both closely associated with the alchemical quest (see Commentary to pp. 47.1–4; 67.21). The tone of his extended satirical attack on alchemy (pp. 44–47) is both playful and earnest, since he was himself committed to the application of chemical research to medicine rather than "multiplying" gold. An interesting analogue is to be found in a work by the alchemical author Michael Maier, called *Lusus Serius* (Oppenheim, 1616),[10] in which representatives of the four-footed creatures, birds, fishes, insects, creeping things, vegetables and minerals, respectively, deliver speeches claiming "superiority under man." The silkworm's claim is followed by those of flax and mercury, and mercury is crowned because of its supposed alchemical virtues.

Like two other contemporary English mock encomia, Sir John Harington's *The Metamorphosis of Ajax* (1596: in commendation of the water-closet)[11] and Thomas Nashe's *Lenten Stuffe* (1599: in praise of the red herring), *The Silkewormes* tends to affirm the importance of man's basic needs by a display of extreme sophistication. The entire first Book is sustained by a patently wire-drawn argument about the original purpose of the silkworm's creation, when neither its food (mulberry leaves) was available nor its function (clothing men) required (pp. 21–23). Moffet exhibits his ingenuity and erudition in an elaborate apparatus of classical allusion and fake authority, the invention of impressive and anciently-attested qualities of the mulberry tree (pp. 51–52), and the proliferation of imaginary inferior rivals to the silkworm (p. 68), but all this is designed both to confer dignity on his subject and to render its doctrine memorable. And there is a deeper implication, too. The provocatively fallacious argument is a Lucianic device that Erasmus and his successors developed in an attempt to suggest that the limitations of scholastic logic extend to the point where even palpably false logic can embody truth. Hence the endorsement of that free play of mind that can take into account accidental features of discourse, puns, and apparent irrelevancies, most of which violate the laws of argument.[12] In *The Silkewormes,* the purpose of such specious argumentation is to expose the miraculous in the mundane and the fascination of the minute.

Date

The Silkewormes and their Flies is a minor Elizabethan poem of considerable interest that has suffered more from the attention of Shakespearian source-hunters than from centuries of neglect. In 1931 Douglas Bush effectively countered Margaret L. Farrand's apparent discovery, in Moffet's version of the story of Pyramus and Thisbe (pp. 9–19), of "An Additional Source for *A Midsummer-Night's Dream*." Unaware of this interchange, A. S. T. Fisher put forward similar arguments for *The Silkewormes* as a Shakespearian source in 1949, and convinced both Kenneth Muir and, in due course, Harold F. Brooks.[13] Though neither Geoffrey Bullough in *Narrative and Dramatic Sources of Shakespeare* (1966–75),[14] nor R. A. Foakes in the new Cambridge edition of the play (1984), has accepted the attribution, they concur in the damning critical assessment of the poem. Dissenting opinions have gone unheeded: scholars like Dwight L. Durling and Kitty Scoular, who rightly consider the poem within the contexts of georgic and nature poetry respectively, accord it some significance, but their brief comments have not affected the Shakespearians.[15] The result is that the student of literary history, instead of approaching *The Silkewormes* as a poem written by a distinguished medical man whose entry in the *Dictionary of National Biography* extends to five columns and whose scientific works attract increasing interest from historians, is likely to assume that as poetry it can be dismissed on the evidence of a relatively short section that at first sight might appear to have provided material for Shakespeare to parody.

The case for *The Silkewormes* as one of Shakespeare's sources, however, is extremely slight and cannot stand up to close examination. The chronological difficulties can only be circumvented by supposing that Shakespeare read the poem in manuscript before he wrote *A Midsummer Night's Dream*. There is no evidence of any such manuscript. An entry in the *Stationers' Register* of 15 January 1589, recording "An Epitaphe or epigram or elegies Done by Master Morfet," can hardly refer to *The Silkewormes* but, if Moffet is the author, probably to his Latin elegy for Sir Philip Sidney, *Lessus Lugubris*.[16] All the evidence in *The Silkewormes* points to a date of composition in the late 1590s: the references to the literary activities of Mary Herbert, Robert Sidney and Philip Sidney's daughter Elizabeth (p. 1 and the dedicatory verses) and the commendation of sericulture to Queen Elizabeth, who was expected at Wilton in 1599.[17] Contemporary efforts to establish sericulture in France, England, and Virginia all depended on royal patronage, so it is likely that the poem was written for such an occasion as a royal visit. One further witness opposed to an early date for *The Silkewormes* is the manuscript of Moffet's most famous work, the *Insectorum theatrum* (BL MS. Sloane 4014). Originally completed in 1590, this manuscript contains numerous later additions, including a reminder "to putt in the treatise of Hieronymus Vida, who hath written the best of all other" (fol. 168b). It is quite likely that Moffet read Vida's *De bombyce* in the late 1590s for the purpose of revising the *Theatrum* and was inspired to write *The Silkewormes*.[18]

The stylistic features of the poem render the likelihood of a Shakespearian parody even more remote. There are a few minor verbal duplications in *A Midsummer Night's Dream*, but these scarcely merit consideration, indicating, as Douglas Bush puts it, no more than that "the two authors were treating the same subject and writing Elizabethan English."[19] The further contention that Moffet's style is bombastic and naive, laying itself open to parody, is no more convincing. The target of Shakespeare's humor is the old-fashioned dramatic style, and he would have gained little by alluding to part of a poem he and a few others may have read in manuscript. Though much of Moffet's versifying is undistinguished Spenserian imitation (see especially pp. 29–31), he wrote in a high style thought entirely appropriate to a scientific poem, both by the authority of Puttenham,[20] and the theory and practice of du Bartas, from whom he derived his penchant for neologisms, circumlocution and mythological allusion.[21] Critics, such as R. A. Foakes, who find Moffet's "blindness" to the implications of his language "astonishing,"[22] have also overlooked the fact that *The Silkewormes and their Flies* is often intentionally comic, employing *ottava rima* for mock heroic effect after the manner of Spenser's *Virgils Gnat* and *Muiopotmos,* and inheriting many of the devices of the Lucianic mock encomium. It is marked by verbal inventiveness[23] and flexibility of tone, ranging from the epigrammatic to the hortatory and the openly satiric.

Authorship

Thomas Moffet's authorship,[24] first suggested by John Chamberlain in 1599, has never been seriously challenged. On 1 March of that year Chamberlain sent a copy to Dudley Carleton, explaining, "The Silkeworme is thought to be Dr. Muffets and in mine opinion is no bad peece of Poetrie."[25] Nathaniel Baxter, in *Sir Philip Sydneys Ourania* (1606), declared of the silkworm:

> Her forme, her life, her foode, her worke, her end,
> By Doctor *Muffet* is eloquently pen'd

and noted in the margin, "Doct. *Muffets* Booke of the Silke-worme" (sig. G3ᵛ).

Moffet (1553–1604) was educated at Cambridge, at both Trinity and Caius Colleges, taking his B.A. in 1573 in the same class as Edmund Spenser.[26] He studied medicine at Basle, and became the leading proponent of Paracelsian medicine in England after his return in 1580.[27] During a fashionable and successful medical career he was personal physician to both Peregrine Bertie, Lord Willoughby and Robert Devereux, the Earl of Essex. His most important connection from a literary point of view was with the Countess of Pembroke. He had attended her brother Sir Philip Sidney and his wife Frances Walsingham in 1585, and moved to Wilton as a pensioner of the Earl of Pembroke in 1593. In that year he wrote an exemplary life of Philip Sidney, *Nobilis,* presented as a New Year's gift to the boy William Herbert.[28] His other abiding interest was entomology, and he was occupied for many years with preparing the papers of his friend Thomas Penny (d. 1588). These entomological studies were derived in turn from the papers of Edward Wotton and Conrad Gesner, and eventually appeared, under Moffet's name, as the *Insectorum theatrum* in 1634, through the efforts of Theodore de Mayerne.[29] This treatise was for many years the standard European work on the subject.

Text

The Silkewormes and their Flies is a very fine example of late Elizabethan literary book production in general and the craftsmanship of Valentine Simmes in particular. With its strikingly attractive title-page (ornamented with woodcuts of the poem's subject), its clear, uncluttered, accurate text, wide margins and neatly disposed sidenotes, it presents an impressively clean and simple appearance. If Moffet subvened the production, it was certainly a work he could be proud of, and its physical appearance (no doubt deliberately) belies the pretended authorship, "T. M. *a Countrie Farmar, and an ap*-prentice in Physicke."

Of the purpose of publication it is hard to be certain. From John Chamberlain's remarks to Dudley Carleton (see above, p. XVIII) it appears that the poem was offered anonymously to provoke speculation: "The Silkeworme is thought to be Dr. Muffets." Two identifiable owners of copies of the poem were Thomas Clayton, later Professor of Music at Gresham College (1607–10), and John Robinson, a Cambridge graduate and later pastor to the pilgrim fathers. Given Moffet's medical and scientific connections, we must suppose that he expected the poem to be welcome in both courtly and learned circles. That it was thus intended as a source of entertainment and admiration for its curious subject, erudition, and pedigree (Vergil and Vida) is not inconsistent with a genuine desire to promote sericulture.

The poem (STC 17994) was one of three known works printed by Valentine Simmes for Nicholas Ling in 1599.[30] The other two were Greene's *Menaphon* (STC 12273) and Nashe's *Lenten Stuffe* (STC 18370). The latter, subtitled "The Prayse of the Red Herring," has some literary affinity with *The Silkewormes,* but was not accorded the same care in point of presentation, the printing being shared with Thomas Jordan (sigs. B-F). Among Ling's other publications for the year was an issue of Drayton's *Englands Heroicall Epistles* (STC 7195), an octavo printed by J. Roberts.[31] In 1603 Simmes was to print, for Ling, the first quarto of *Hamlet* (STC 22275), and he also printed works by Chapman, Drayton, Marston and Daniel. Moffet's only known English poem was thus handled by stationers with established literary connections. Appropriately, Simmes was to set himself up in "White Friars neere the Mulberry Tree" (1610), though in 1599 he was still presumably "dwelling on Adling hil at the signe of the white Swanne."[32]

Bibliographical Description

Title page: See facsimile.

Collational Formula:

  4°: A-K^{4}L^2 [3 signed (–A2, A3, L2)] 42 leaves:
  pp. *i–viii* 1–23 *24* 25–75 *76* = 84 pages

  RT] *Of the Silke wormes | and their Flies.* [silke wormes, sigs. B3^v, D3^v, D4^v, F3^v, F4^v, H3^v, H4^v, K4^v]

  Inconsistent catchword: K4^v "Large" for "Lay."

Contents:

A1, blank except for the signature, A; A1ᵛ, blank; A2, title-page; A2ᵛ, blank; A3, dedicatory verses; A3ᵛ–A4, "The Table"; A4ᵛ, "Faults escaped in Printing"; B1–F4ᵛ, Book I; G1–L2, Book II; L2ᵛ, blank.

Notes on the Printing:

Although the title page illustration, consisting of three woodcuts surrounded by a border of printer's flowers, does not correspond exactly to any of the illustrations in the printed edition of Moffet's *Insectorum theatrum* (1634), it may have been connected with a projected edition of that larger entomological work. It was in about 1603–04 that William Rogers was commissioned to prepare an engraved title-page for the *Theatrum,* when it was re-dedicated to King James.[33] In the manuscript of the *Theatrum* (BL MS. Sloane 4014), there are, unusually, no illustrations for the relevant chapter, *"De ortu, generatione, alimento et Metamorphosi Erucarum"*; perhaps they had been used to prepare the title page for the poem.

The ornament that connects the printing of *The Silkewormes* most surely with Valentine Simmes is that which appears on signature A4ᵛ ("Faults escaped"). It portrays a cherub with chaplet and ribbon and was used by Simmes in the printing of Annibale Romei's *The courtiers academie* (1598; STC 21311) and John Weever's *The mirror of martyrs* (1601; STC 25226).[34] Two types of printer's flowers commonly used by Simmes are attractively employed, one on the title page and the other for rules at the end of "The Table" and the beginning of each Book of the poem. The type-face used in the setting of the main text is identified by Ferguson (p. 42) as 93 Roman; it is almost certainly the same as that described by Frank Isaac as Simmes's 94, a Garamond St. Augustine face "with a narrower set."[35] The measure used for the main text is 87 mm (as in seven other books printed by Simmes), with a sidenote column measuring 17.5 mm.

The paper used for the 1599 quarto bears three watermarks, which can be identified with Ferguson's tracings, serial numbers 32 (a fleur-de-lis pattern), 37 (a star pattern) and 38 (a trifoil above the letter "M"; Ferguson, pp. 67–68, 76). Simmes drew on the same stock of paper for the printing of the Shakespeare quartos of *Much adoe about Nothing* (37 and 38) and *The second part of Henrie the fourth* (32 and 38) in 1600. In the various copies of *The Silkewormes* that I have collated there is no discernible pattern in the occurrence of the watermarks.

Analysis of the running titles shows that Simmes used a pair of skeletons, one, characterized by lower-case "s" in *"silke wormes,"* being used for both the inner and outer formes of alternate sheets. This pattern also occurs in Simmes's printing of John Weever's *Faunus and Melliflora* (1600; STC 25225). But the same broken letter " ɔ " appears in connection with both skeletons in our text (p. 36.2 on F2ᵛ, and p. 71.6 on K4). There is some evidence that two compositors were involved in the setting, in that the numerals that refer the reader from the text to the sidenotes appear without parentheses in signatures B to C and the outer formes of signatures D to E, whereas parentheses appear in signature F to the end and in the inner formes of D to E (except D4).[36] Both styles, with and without parentheses, are to be found on sig. D3ᵛ (p. 22) and F4 (p. 39). It is notable that all printing errors connected with these numerals occur in the formes in which parentheses are used. A spelling analysis does not, however, indicate any significant divergence between these two divisions of the printed text, nor do these divisions correlate with Moffet's own spelling preferences as evidenced by the only holograph English manuscript, the 1585 letter to Michael Hicks (BL MS. Lansdowne 107 art. 13): neither compositor, if there were two, seems to have been following copy more closely. Moffet's favored spellings "freind" and "ow" (in such words as "owt" and "yow") are nowhere reflected in the 1599 quarto, and though he clearly prefers the ending "es" above "esse," the quarto heavily favors "esse." Both Moffet and Simmes's compositor(s) prefer the "ll" ending and "ie" to "y," although Moffet's preference is more pronounced in both cases.

The form VV is used interchangeably with W throughout the poem, though it is notable that in signatures D and G, VV is confined to the inner formes. The ampersand is normally used to avoid a turn-under line in the text of the poem itself. Italics, normally employed for proper names, appear more liberally in Book II of the poem, being attracted by such words as *Emerode* (p. 49.11), *Hemisphere* (p. 57.24), *Claricalls* (pp. 73.24) and the names of various birds (pp. 43.11; 44.3, 4, etc.), but this feature is probably to be attributed to the copy.

Copies Collated

The revised STC records copies of the quarto at the British Library; the Dyce Collection of the Victoria and Albert Museum; the Bodleian Library; Emmanual College, Cambridge; Edinburgh University; Folger Shakespeare Library; Huntington Library; Boston Public Library; University of Chicago; and Yale University. Of these I have collated four copies:

1. *Bodl^1* = The Clayton copy. Bodleian Library, pressmark Art. 4° M.2 BS. This is bound in vellum with leather ties, and is inscribed "Thomas Clayton." The errata have been entered in the text, somewhat inaccurately, in a contemporary hand, the same as that of the fly-leaf inscription. There is also one original emendation, "springs" (p. 3.20), which I have adopted.

2. *Bodl^2* = The Douce copy. Bodleian Library, pressmark Douce M.404. This has been re-bound handsomely in leather and contains two inserted leaves of miscellaneous notes on the poem. Of the 1599 edition, only the leaf A4, gatherings B to I, and the first three leaves of gathering K, are included, the remainder being meticulously supplied in facsimile by hand.

3. *HEH* = The Huntington Library copy, pressmark 62600. This contains the bookplate of Robert Hoe.

4. *BL* = The Robinson copy. British Library, pressmark 239 k. 19. The title page bears the inscription "Jo: Robinson"; a later owner was William Hall, the eighteenth-century poet and antiquary, who rebound the book in leather. Some of the printer's errata list have been entered, and there is a pencilled marginal comment on p.1.17, "sweete heav'ns restore that time," referring us to Ophelia (presumably her cry, "O heavenly powers restore him").

I have also examined a copy at the Boston Public Library (pressmark G 389 b.3 = *BPL*) and two copies at the Folger Shakespeare Library (CS 713 and CS 10). Press variants have been observed in two formes, C inner and G outer:

C2, note 4, line 3.
Corrected: Babilon Bodl¹, BL
Uncorrected: Balilon Bodl², HEH

G4ᵛ lines 1 and 2.
Corrected: But damsals ... Ladyes (I) foe BL, Bodl¹, Bodl², HEH
Uncorrected: But (1) damsels ... Ladyes? foe

Note on the Facsimile

The copy of *The Silkewormes* reproduced here in facsimile is in the Henry E. Huntington Library. Although the Renaissance English Text Society does not normally publish facsimile editions, Simmes's quarto, which contains few printing errors, is of sufficient bibliographical interest, especially in the use of sidenotes, to warrant exceptional treatment.

The text is supported by a conservative critical apparatus, recording editorial emendation, press variants, and the printer's errata list ("Faults escaped"). Where necessary, I have discussed the few substantive emendations in the general commentary. I have almost always resisted the temptation to regularize the use of question marks, many of which anticipate the end of the sentence, as in:

Ofsprings of egges, what are you but a voice?
Angring sometimes your friends with too much noyse.
(p. 43.15–16)

This pattern is repeated so often, without obscuring the sense, that it cannot be regarded as a printer's error.

In quotations from the text in the introduction and commentary, long *s*, *i* and *j*, *u*, *v* and *vv* have been modernized: digraphs are treated as two letters, and contractions, including the ampersand, are expanded.

Notes to the Introduction

1. See Joan Thirsk, "The Fantastical Folly of Fashion: the English Stocking Knitting Industry, 1500–1700," in *Textile History and Economic History: Essays in Honour of Miss Julia de Lacy Mann*, ed. N. B. Harte and K. G. Ponting (Manchester, 1973) 50–73, and *Economic Policy and Projects* (Oxford, 1978).
2. *Minerva Britanna* (1612) 89.
3. "Preface to the Reader," *The feminine monarchie* (1609).
4. B. H. Newdigate, *Michael Drayton and his Circle* (1961) 155.
5. Line references are to the Latin text printed with an English translation by Samuel Pullein (Dublin, 1750).
6. Sidney's translation was entered in the *Stationers' Register* for August 1588, and is mentioned by Fulke Greville, Florio and Sylvester. See William Ringler, ed., *Poems of Sir Philip Sidney* (Oxford English Texts, 1961) 339, and Susan Snyder, ed., *The Divine Weeks and Works of Guillaume de Saluste, Sieur du Bartas*, trans. J. Sylvester (Oxford English Texts, 1979) 1: 70.
7. James Carscallen, *English Translators and Admirers of du Bartas, 1578–1625*, (B. Litt. thesis, Oxford University, 1958) 250–52.
8. K. Scoular, *Natural Magic* (1965) 49, introducing selections from the poem, pp. 49–67; on the authorship, see Leslie Hotson, "Marigold of the Poets," *Transactions of the Royal Society of Literature*, n.s., 17 (1938): 47–68. In the S.T.C. this work is listed under Thomas Cutwode. It was probably in circulation by 1595; see R. Krueger, ed., *The Poems of John Davies* (Oxford English Texts, 1975), commentary to *Orchestra*, stanza 129.
9. See Moffet's *The Theater of Insects* (1658 trans.) 932.
10. J. de Salle, trans. (1654) as *Lusus serius: or serious passe-time. A philosophicall discourse concerning the superiority of creatures under man*.
11. Moffet refers to this in *Healths Improvement* (1655) 13–14.
12. See W. J. Kaiser, *Praisers of Folly* (1964), esp. 1–16; Joel B. Altman, *The Tudor Play of Mind* (Berkeley, 1978); and A.E. Malloch, "The Techniques and Function of the Renaissance Paradox," *Studies in Philology* 53 (1956): 191–203.
13. Farrand, in *Studies in Philology* 27 (1930): 233–43; Douglas Bush, "The Tedious Brief Scene of Pyramus and Thisbe," *Modern Language Notes* 46 (1931): 144–47; Fisher, "The Source of Shakespeare's Interlude of Pyramus and Thisbe: A Neglected Poem," *N & Q* 194 (1949): 376–79, 400–402; Muir, "Pyramus and Thisbe: A Study in Shakespeare's Method," *Shakespeare Quar-*

terly 5 (1954): 141–53, and *The Sources of Shakespeare's Plays* (1977) 73; Brooks, ed., *A Midsummer Night's Dream* (New Arden ed., 1979) lxiv.

14. Vol. 1 (1966) 375.

15. Durling, *Georgic Tradition in English Poetry* (New York, 1935) 5, 33; Scoular published generous selections of the poem to introduce her chapter "Much in Little" in *Natural Magic* (1965). See also John Buxton, *Sir Philip Sidney and the English Renaissance* (1954) 233; Katherine Duncan-Jones, "Pyramus and Thisbe: Shakespeare's Debt to Moffett Cancelled," *RES* 32 (1981): 296–301; and Michael G. Brennan, *Literary Patronage in the English Renaissance* (1988), 76–77.

16. This poem is contained in the manuscript of *Nobilis* (Huntington Library MS. HM 1337).

17. Duncan-Jones, 298; Mary Herbert appears to have prepared a presentation copy of her translation of the Psalms, but the visit did not take place, probably because both the Queen and Henry Herbert were ill. See Brennan, *Literary Patronage*, p. 96.

18. One of the MS. changes in the *Theatrum* is the alteration of the dedication from Elizabeth to James.

19. Bush, 145.

20. *The Arte of English Poesie*, ed. G. D. Willcock and A. Walker (1936) 44.

21. See Susan Snyder, ed., *The Divine Weeks and Works* (trans. Sylvester) 1: 53.

22. R. A. Foakes, ed., *A Midsummer Night's Dream* (Cambridge, 1984) 11.

23. It provides the *Oxford English Dictionary* with almost twenty first citations (see Glossary).

24. The name is variously spelled Moffet, Moffett, Moufet, and Muffet; I have adopted the form most frequently used in the English manuscript evidence: see the family pedigree, Bodleian MS. Ashmole 799, fol. 130; letter to Essex, *H.M.C.* Hatfield House IV, 174; letter to Michael Hicks, BL MS. Lansdowne 107 art. 13; the manuscripts of *Nobilis,* Huntington Library MS. HM 1337, and the *Insectorum theatrum,* BL MS. Sloane 4014, name him Moffetus.

25. N. E. McClure, ed., *Letters of John Chamberlain* (Philadelphia, 1939) 1: 70.

26. J. R. Tanner, *The Historical Register of the University of Cambridge* (1917) 384; Moffet took his M.A. in 1576.

27. As a medical student in Basle he ran into trouble for attacking the anti-Paracelsian Erastus in his doctoral thesis, *De anodinis medicamentis eorumque causis et usibus* (Basle, 1578); after returning to England he published an influential "apologetic dialogue" *De iure et praestantia chymicorum medica-*

mentorum (Frankfurt, 1584) that challenged the insouciance of the [Royal] College of Physicians. Later he participated in a project to set up a London Pharmacopeia that included chemical remedies for internal use.

28. *Nobilis* was first published in 1941 in an edition by V. B. Heltzel and H. H. Hudson (San Marino, California), which contains an English translation and an account of the author's life.

29. There is a fine English translation by John Rowland, appended to his edition of Edward Topsell's *The history of four-footed beasts and serpents* (1658).

30. For Valentine Simmes's career, see W. Craig Ferguson, *Valentine Simmes* (Charlottesville, Virginia, 1968); cited as Ferguson.

31. Paul G. Morrison, *Index of Printers, Publishers and Booksellers* (Charlottesville, Virginia, 1961), *sub* Ling.

32. R. B. McKerrow, ed., *A Dictionary of Printers and Booksellers in England, Scotland and Ireland, and of Foreign Printers of English Books, 1557–1640* (1910) 245.

33. Arthur M. Hind, *Engraving in England in the Sixteenth & Seventeenth Centuries* (1952) 1: 276–77, and Plate 152.

34. Ferguson 47, no. 14.

35. *English Printers' Types of the Sixteenth Century* (1936) 39, and Plate 68.

36. I am indebted to Professor A. R. Braunmuller for pointing this out to me.

THE
Silkewormes, and their Flies:

Liuely described in verse, by T. M.
a *Countrie Farmar*, and an apprentice in Physicke.

For the great benefit and enriching of England.

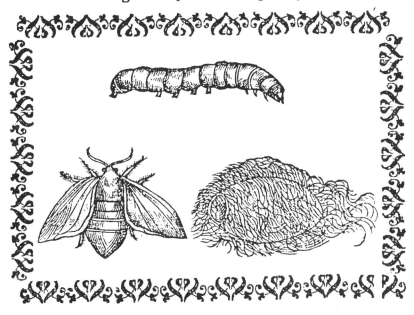

Printed at London by V. S. for Nicholas Ling, and are to be sold at his shop at the West ende of Paules. 1599.

To the moſt renowned Patroneſſe, and noble *Nurſe of Learning* MARIE Counteſſe of Penbrooke.

G*Reat enuies Obiect, Worth & Wiſedoms pride,*
Natures delight, Arcadia's *heire moſt fitte,*
Vouchſafe a while to lay thy taske aſide,
Let Petrarke *ſleep, giue reſt to* Sacred Writte,
Or bowe, or ſtring will breake, if euer tied,
Some little pawſe aideth the quickeſt witte:
  Nay, heau'ns themſelues (though keeping ſtil their way)
  Retrogradate, and make a kind of ſtay.

I neither ſing Achilles *baneful ire,*
Nor Man, nor Armes, nor Belly-brothers warres,
Nor Britaine *broiles, nor citties drownd in fire,*
Nor Hectors *wounds, nor* Diomedes *skarres,*
Ceaſe country Muſe *ſo highly to aſpire:*
Our Plaine *beholds but cannot holde ſuch ſtarres:*
  I*oue-loued wittes may write of what they will,*
  But meaner Theams beſeeme a Farmers quill.

I ſing of little Wormes and tender Flies,
Creeping along, or basking on the ground,
Grac't once with thoſe thy heau'nly-humane eies,
Which neuer yet on meaneſt ſcholler fround:
And able are this worke to æterniſe,
From Eaſt to Weſt about this lower Round,
  Deigne thou but breathe a ſparke or little flame
  Of likeing, to enliſe for aye the ſame.

          Your H. euer moſt bounden.
                   T. M.

The Table.

1 WHen garments were first vsed. Fol. 2
2 Whereof garments were first made. 3
3 Diuers opinions how and when silke was first inuented and worne. 4 5 6 7 8 9 10 11 12 13 14 15 16 17 18 19 20.
4 Whether the Silke-worme or the Silke Flie were first created. 21 22 23.
5 Whether the egge or the henne be first in nature. 24 25 26 27.
6 Why the silke flies are onely of a white colour. 28.
7 What day of the weeke they were first created. 29 30 31.
8 The rare vertues and chastity of Silke Flies. 27 28 33 38
9 That they are not to be scornd as being little and therefore contemned creatures. 34 35 36 37.
10 Their wonderful encrease and propagation. 39 and 65 66.
11 Their manner of dying. 40 66.
12 That their egges are more worth then the egges of any Flier: yea then the Philosophers egge, if there were any such. 41 42 43 44 45 46 47.
13 Their egges must be kept in a temperate place. 48.
14 They are not to be hatched til the Mulbery tree buds. 50.
15 Why the Silke-wormes beeing crept out of the shel feede onely vpon Mulbery leaues. 51 52 53.
16 How long they feede: When their meate is to be gathered; In what quantity they are to be dieted. 55.
17 That variety of meates is naught for them. 56.
18 Their table is to be kept cleane. ibid.
   How the sicke are discerned. 59.
19 Of their sleepe: ibid.
20 How they must be distributed when they grow great. 57.
21 The causes of their sicknesse. 59.
22 Signes

The Table.

22 Signes of their readinesse to worke, and how then they must be vsed. 60.
23 How long they worke. 61.
24 When the wormes are metamorphosed into Flies. 61.
25 When and how their silke is to be winded vp. 61.
26 The sorts and vse of their silken threede. 63 64.
27 How their egges are to be preserued. 67.
28 That the silke of Silke-wormes is the best of al other. 68.
29 What profit and pleasure ariseth in keeping of them. 71 72 73 74.
30 Keeping of Silke-wormes hindereth neither Shepheards, Spinsters, Weauers, nor Clothiers. 75.

FINIS.

Faults escaped in Printing.

page	line	fault		correction
5	11	the		thy
7	3	euer		neuer
7	14	courſer		Courſers
9	19	priuate		Priuie
17	3	his		this
17	13	layes		laye
19	2	for h harſes	reade	herſes
27	4	through		thorough
29	10	through		thorough
48	2	Enicthean		Erycthean
56	1	us		as
59	1	I any		If any
66	15	drope		drop

Of the Silke wormes and their Flies.

Ydneian Muse: if so thou yet remaine,
In brothers bowels, or in daughters breast,
Or art bequeath'd the *Lady of the plaine*,
Because for her thou art the fittest guest:
Whose worth to shew, no mortall can attaine,
Which with like worth is not himselfe possest:
   Come help me sing these flocks as white as milke,
   That make, and spinne, and die, and windle silke.

For sure I know thy knowledge doth perceiue,
What breth embreath'd these almost thingles things:
What Artist taught their feete to spinne and weaue:
What workman made their slime a robe for kings,
How flies breed wormes, how wormes do flies con-
Frō natures womb, how such a nature springs, (ceiue:
   Whereof none can directly tell or reede,
   Whether were first, the flie, the worme, or seede.

A time there was (sweete heau'ns restore that time,)
When bodyes pure to spotlesse soules first knit,
Deuoyd of guilt, and ignorant of crime,
Vpright in conscience, and of harmelesse wit,
Disdaind to weare a garment nere so fine,
As deeming coates and couers most vnfit,
   Where nothing eie could see, or finger touch,
   Which God himselfe did not for good auouch. *Gen.1.verse 31.*

Yea

Yea, when all other creatures looked base,
As mindful onely of their earthly foode:
Or else as trembling to behold the place,
Where iudge eternall sate, and Angels stood:
Then humane eyes beheld him face to face,
And cheekes vnstain'd with fumes of guiltie bloud,
  Desir'd no maske to hide their blushing balles,
  But boldly gaz'd and pried on heau'nly walles.

The breast which yet had hatcht no badde conceat,
Nor harbor'd ought in heart that God displeaz'd,
Did it for silken wastcotes then intreate?
Sought it with Tyrian silks to be appeaz'd?
No, no, there was no neede of such a feate,
Where all was sound, and members none diseaz'd:
  Nay more, The basest parts and seates of shame,
  Were seemely then, and had a comely name.

Gen. 3.   But when selfe-will and subtile creepers guile,
Made man to lust, and taste what God forbad,
Then seem'd we to our selues so foule and vile,
That straight we wisht our bodies to be clad,
Seeing without, and in such great defile,
As reft our wittes, and made vs al so mad:
  That we resembled melancholique hares,
  Or startling stagges, whom euerie shadow scares.
                                        Then

Then Bedlam-like to woods wee ranne apace,
Praying each tree to lend vs shade or leaues,
Wherewith to hide (if ought might hide) our face
From his al-seeing eyes, who al perceaues,
And with ful-brandisht sword pursues the chace,
Traitors of rest, of shade, and al bereaues:
 Permitting men with nothing to be clad,
 But shame, dispaire, guilt, feare, and horror sad.

These robes our parents first were deckt withal,
Then figtree fannes vppon their shame they wore:
Next, skinnes of beasts, (to shew their beastly fall)
Then, hairy cloathes, and wooll from Baa-lambs tore,
Last, Easterne wittes, from mane of Camels tall, *Plin. lib. 12. ca.*
Made water-waued stuffe vnseene before, *10. & lib. 24.*
 But til the floud had sinners swept away, *cap. 12.*
 Nor Flaxe, nor Silke, did sinful man array.

For so it seemed iust to Iustice eyen,
Defiled men to weare polluted things:
And Rebels not to clothe in Flaxe or line,
Which from the sacred loines of *Vesta* spring,
Cleane, knotlesse, straight, spotlesse, vpright, and fine,
VVhose floure is like fiue heau'nly-azurd wings,
 Whose slime is salue, whose seed is holsom food,
 whose rinde is cloth, whose stuble seru's for wood *Plut. lib. de Iside*
         Or *& Osir.*

Of the Silke wormes

¹ *A most famous spinner in Lydia, of whom Ouid 6 metam.*

Or if 1 *Arachne* erst made sisters threed,
Was it thinke you, for euery man to weare?
Or onely for the sacrificers weede,
VVho of th'immortall priest a type did beare?
Wearing not aught that sprang from brutish seed,
But what from out it selfe the earth did reare:
Exodus 28.   So that till holy priesthood first began,
VVe neuer reade that linnen clothed man.

² *Orpheus a most famous Poet. Ouid 11 met.*

Yet some conceiue when 2 *Theban* singer wanne,
VVood-wandring wights to good and ciuill life,
(Which erst with beares and wolues in desarts ran,
Knowing no name of God, law, house, or wife)
That then his brother *Linus* first began
The Flaxmans craft (a secret then vnrife)
   Deuising beetles, hackels, wheeles, and frame,
   Wherwith to bruse, touse, spin & weaue the same.

But Silke (whereon my louing Muze now stands)
Was it the offspring of our shallow braine?
Spunne with these fingers foule? these filthy hands,
Tainted with bloud, reuenge, and wrongful gaine?
Ah no, who made and numbreth all the sands,
Wil teach vs soone that fancie to be vaine:
   Farre be it from our thoughts, that sinfull sence,
   Should make a thing of so great excellence.

Ne

and their Flies.

Ne neede wee yet with *Tuscane* Prelate flie,[1]
To fictions strange, or wanton *Venus* eyen:
Who seeing *Pallas* taught from *Saturne* hie,
To clothe her selfe and hers with weaued line,
Yea all the Nimphs and Goddesses in skie,
To weare long stoles of Lawne and Cambrick fine:
   Fretted to see her selfe and boy new borne,
   Left both to heau'n and earth an open scorne.

[1] *Hironimus Vida, Bishop of Alba. lib.1. de Bombyce.*

Reuenge she cri'de vnto the sire of *Ioue*,
As she lay hidde vnder th'Idalian tree:
Affoord some rayment from the house aboue,
If but to hide the shame of mine and mee.
So may thou learne from vs *The art of Loue*,
Whereby to winne each Ladies heart to thee.
   But grumbling Chuff reiected still her prayre,
   Whereat lamented heau'ns and weeping aire.

Then Cyprian Queene perceiuing that no cries
Could pierce the leaden eares of sullen Sire,
Straight lodg'd her sonne in faire *Phillyraes* eies,[2]
And caus'd him thence to darte vppe such a fire,
As had consum'd the very starres and skies,
Yea melted *Saturnes* wheeles with hot desire:
   Vnlesse that very houre he had come downe,
   And beg'd her aide, on whom he late did frowne.

[2] *Oceanus his daughter, a most braue virgin. Ouid 6 met.*

How

How often, as his loue on *Pelion* hill
Stoopt downe to gather herbs for wounds and sores,
Strew'd he before her Tutsan, Balme, and Dill,
Long Plantaine, Hysope, Sage, and Comfrey moares?
Offring besides, the art and perfect skill,
Of healing bloudy wounds and festred coares:
   How oft (I say) did he each day descend,
   And booteleffe al his vowes and wooings spend?

He lou'd, she loath'd, he liked, she disdain'd:
He came, she turn'd, he prest, she ranne away,
Neither by words, nor gifts shee could be gain'd,
(For onely in her eies the Archer lay)
Regarding nought but (wherein she was train'd)
Wounds how to cure, and smartings to allay:
   As for the wound of Loue, she felt it none,
   And therefore litle heeded *Saturns* mone.

Thus thus perplext the chiefe and grauest God,
(Or rather God suposd of highest place)
Toucht now, nay throughly scourg'd with *Cupids* (rodde,
Sent from the eyes but of a mortal face,
Flewe downe forthwith where *Venus* made abode,
And prostrate lying at her feete for grace:
   Promis'd the richest clothing for her Art,
   That now she did, or could desire in hart.

Who

and their Flies. 7

Who careleffe of reuenge, and innely grieu'd,
(True beauty aye is ful of rueful mone)
Was euer wel til *Saturne* was releeu'd,
His inward griefes aſſwag'd, & ſorrowes gone.
And finding him, of hope, and helpe, bereeu'd,
(For ſtill *Phillira* was more hard then ſtone)
    Sith that, quoth ſhe, the virgin ſcorns thy loue,
    Try whether craft and force wil make her moue.

Transforme thy ſelfe into a Courſer braue,
(What cannot loue transforme it ſelfe into?)
Feede in her walkes: and in a moment haue
What thou haſt woo'd to haue with much adooe:
Whereto, conſent the auncient Suter gaue,
In courſer clothes, learning a maide to wooe,
    Filling ech wood with neighs and wihyes ſhrill,
    Whilſt he poſſeſt his loue againſt her will.

For leſſon which, his Miſtris to requite,
Not with vaine hopes in lieu of friendly deeds,   *1 Mercurie,*
By *Maiae's* 1 ſonne (before it grew to night)   *poſtmaſter to*
He ſent a Napkin ful of little ſeeds,   *Iupiter.*
Tane from the tree where *Thisbes* ſoule did light,
To make her ſelfe and boy farre brauer weeds,
    Than *Pallas* had, or any of the ſeu'n,
    Yea, then proud *Iuno* ware the Queene of heau'n.
                              Withal

Withall, by him he sent the mysterie
Of weauing silke, which he himselfe had found,
When chac'd from heau'n by sonnes owne trechery,
Hee was compel'd to wander here on ground,
Where, in the depth of griefe and pouertie,
The heigth and depth of Arts he first did sound:
   Yet would he this to none but her reueale,
   By whose deuise hee did *Phillyra* steale.

What? shall we thinke, that silke was a reward.
Bestow'd on craftie dame for aide vniust?
Would men, nay, ought they haue such hie regard,
Of that which was the lone and hire of lust?
Not so, what ere th'Italian Bishop dar'd
To faine for true, and giue it out with trust:
   Yet sith silke robes the blessed High-priest wore,
   They were not sure the first fruits of a whore.

<small>1
Plinius Secundus, lib. 11. *cap.* 2
2
Called *Pamphila, a most princely Damsell.*</small> *Vespasians* 1 Scribe affirmes in *Cean* Ile,
  Latous 2 daughter, quicke of eye and wit,
  Hunting abroad, times trauaile to beguile,
  Chaunc'd at the length vnder a tree to sitte,
  Where many silken bottoms hangd in piles,
  One by another plac't in order fit.
    Shee tooke one downe, and with her faulcon eye,
    Found out the end that did the rest vntie.
                                   Looke

and their Flies.

Looke how the hungry Lambe doth friske and play,
With restlesse taile, and head, and euery limbe,
When it hath met his mother gone astray,
Who absent blear'd and tear'd as much for him:
Or as *Aurora* leapes at breake of day,
Seeing her louely brother rise so trim,
   No lesse that Princesse triumph't(if not more)
   Finding out that which was not found before.

Loues Schoolemaster 1 records a tale most sweete,   1 *Ouid lib.4.*
Of louers two that dwelt at *Babilon*,   *Metam.*
Equall of age, in worth and beautie meete,
Each of their sex the floure and paragon,
Next neighbours borne on side of selfesame streete,
For twixt their parents houses dwelled none,
   Him *Pyramus*, her *Thisbe* men did call,
   Coupled in heart, though seuered by a wall.

As neighbours children, oft they talke and view,
That neighbourship was formost steppe to loue,
Loue, which (like priuate plants) in short time grew,
Pales, wals, and eues, yea houses and all aboue,
Nay Hymeneus feasts were like t'ensue,
And sacred hands giue ring and wedding gloue,
   Had not vnhappie parents that forbad,
   Which to forbid, no cause but wil, they had.

Of the Silke wormes

If louers spake, it was now all by lookes,
None deign'd or durst be trouchman to their mind,
Paper was barr'd, and penne, and inke, and bookes,
Not any helpe these parted prisoners find,
But of a rift along the wal that crookes,
(A wall of flint, yet more then parents, kind)
   Which, were it old or new, none it espies,
   But louers quicke, al-corner-searching eyes,

This rift they vsde, not onely as a glasse,
Wherein to see daily each others face,
But eke through it their voyces hourely passe,
In whispring murmurs with a stealing pace:
Sometimes when they no longer durst (alas) (place,
Send whisprings through, when keepers were in
   Yet would they shift to blow through it a breath,
   Which fed & kept their hoping harts from death.

Enuious wal (sayd they) what wrong is this?
Why doth not loue or pittie make thee fal?
Or (if that be for vs too great a blisse)
Why is thy rift so narrow and so small,
As to deny kind loue a kindly kisse?
For which we neuer proue vnthankful shal,
   Although in truth we owe inough to thee,
   Giuing our eyes and voyce a way so free.

                                                In

and their Flies.

In vaine thus hauing plaind in place distinct,
When night approacht, they ech bad ech adew,
Kisfing their wal apart where it was chinckt,
Whence louely blasts and breathings mainely flew:
But kisses staide on eithers side fast linckt,
Seal'd to the wal with lips and Louers glue:
   For though they were both thick and many eake,
   Yet thicker was the wal that did them breake.

Rose-fingred 1 Dame no sooner had put out   *1 The morning, Homer. Iliad. 4.*
Nights twinckling fires and candles of the skie,
Nor *Phœbus* 2 brought his trampling steeds about,   *2 The Sunne.*
Whose breath dries vp the teares of *Vestaes* 3 eie,   *3 The earth.*
But swift and soft, without all noyse or showt,
To wonted place they hasten secretly,
   Where midst a many words muttred that day,
   Next midnights watch, each vowes to steale away.

And left when hauing house and cittie past,   *4 Which was without the gates of Balilon, towards the forest. Sabell. Enneiad. 1. cap. 6.*
They yet might erre in fields, and neuer meete,
At *Ninus* 4 tombe their *Rendes-vous* is plac't,
Vnder the Mulb'ry white, and hony-sweete:
Growing hard by a spring that ranne at waste,   *5 The swift riuer of Donawe.*
With streames more swift then speedy 5 *Isters* feete.   *6 The Charles waine.*
   There they agreed in spite of spite to stand,   *7 The great star following Vrsa maior.*
   Whē 6 Monarchs teame had past 7 *Bootes* hand.

Of the Silke wormes

Consent they did, and day consented too,
Whose Coach ranne downe the seas in greater hast,
Then euer it was wont before to doo,
Loue-louing night approched eke so fast,
That darknesse leapt, ere twilight seem'd to go,
Wherat though some gods frown'd, some were
    Yet *Lethes* 1 brother did the louers keepe, (agast,
    Chaining their guard with long and heauy sleep.

1 Sleep the brother of forgetfulnesse. Cic. lib. de nat. deorum.

How feately then vnsparred she the doore?
How silent turn'd it on the charmed cheekes?
And being scap't, how glad was she therefore?
How soone arriu'd where she her fellow seekes?
Loue made her bold, loue gaue her swiftnesse more
Then vsually is found in weaker sexe,
  But all in vaine: nay rather to her ill,
  For haste made waste, and speede did speeding kil.

The grisly wife of brutish monarch strong,
With new slaine prey, full panched to the chinne,
Foming out bloud, came ramping there along,
To siluer spring, her thirst to drowne therein,
Whereat the fearefull maide in posting flung,
(For 2 *Lucines* eye bewrayde the Empresse grimme)
  Into a secret caue: and flying, lost
  A scarfe (for *Pyrams* sake) beloued most.

2 The Moone-shine.

               When

and their Flies.

When sauage Queene had wel her thirst delayde,
In cooling streames, and quenched natures fire,
Returning to the place where late she prayde,
To eate the rest when hunger should require,
In peeces tore the scarfe of haplesse maide,
With bloudy teeth, and firie flaming ire,
    Whilst she (poore soule) in caue plaid least in sight,
    Fearing what should her loue befall that night.

Who comming later then by vow he should,
Perceiu'd a Lions footsteps in the sand,
Whereat with face most pale, and heart as cold,
With trembling feare tormented he doth stand.
But when he sawe her scarfe (wel knowne of old)
Embru'd with bloud, and cast on either hand?
    O what a sigh he fetcht? how deepe he gron'd?
    And thus, if thus : yea, thus he inly mon'd.

Shalt thou alone die matelesse, *Thisbe* mine?
Shall not one beast be butcher to vs both?
What? is my *Thisbe* reft of life and shine?
And shal not *Pyram* life and shining loath?
Mine is the cursed soule, the blest is thine,
Thou kep'st thy vow, I falsified mine oath,
    I came too late, thou cam'st (alas) too soone,
    Too dangerous standing, by a doubtfull moone.

O Lions fierce (or if ought fiercer be,
Amongst the heards of woody outlawes fell)
Rent, rent in twaine this thrise-accursed me:
From out your paunch conuey my soule to hell:
Whose murdring flouth, and not the sisters three,
Did *Thisbe* sweete, sweete *Thisbe* fowly quell:
   But cowards onely call & wish for death,
   Whilst valiant hearts in silence banish breath.

Then stooping, straight he took hir scarfe frō ground,
And bare it with him to th'appoynted place,
Kissing it oft, watring each rent and wound,
With thousand teares, that trailing ranne apace.
Salt teares they were, sent from his eyes vnsound,
Yea salter then the sweate of Oceans face:
   At last (hauing vnsheath'd his fatall blade)
   Thus gan he cry, as life beganne to fade.

Hold earth, receiue a draught eke of my bloud,
(And therewith lean'd vppon his sword amaine)
Then falling backward from the crimsin floud,
Which spowted forth with such a noyse and straine,
As water doth, when pipes of lead or wood,
Are goog'd with punch, or cheesill slit in twaine,
   Whistling in th'ayre, & breaking it with blowes,
   Whilst heauie moysture vpward forced flowes.

The Mulb'ry ſtrait(whoſe fruit was erſt as white
As whiteſt Lilly in the fruitfullſt field)
Was then and euer ſince in purple dight,
Yea euen the roote no other ſtaine doth yeeld,
With blackiſh gore being watred all that night,
In morneful ſort, which round about it wheel'd,
    Onely her leaues retaind their former hue,
    As nothing toucht with death of louer true.

No ſooner was hee falne, and falling, freed
Of perfit ſence: but ſhe ſcarce rid of feare,
Returnes againe to ſtanding fore agreed,
Not dreaming that her loue in kenning were,
Her feete, her eyes, her heart and tongue made ſpeed,
To vtter all things lately hapned there,
    And how ſhe ſcap't the Lioneſſes clawes,
    By letting fall a ſcarfe to make her pawſe.

But when ſhe vewd the newly-purpled face
Of Berries white: that changing chang'd her mind,
New ſignes perſwade her, that is not the place,
By either part to meete in fore aſsign'd.
Thus doubting whilſt ſhe ſtood a little ſpace,
She heard a flittering carried with the winde,
    And viewed ſomewhat ſhake in quiu'ring wiſe,
    Which ſtraite reuok't hir feete, but more her eies.
                                     Her

Her lippes grew then more pale then paleſt Boxe,
Her cheekes reſembled Aſhwood newly feld,
Grayneſſe ſurpriz'd her yellow amber locks,
Not any part their liuely luſtre held:
Yea euen her vent'rous heart but faintly knocks,
Now vp, now downe, now falne, now vainly ſweld,
   Toſt like a ſhippe when 1 *Corus* rageth moſt,
   That ankers hath, and maſts and maſter loſt.

1 *One of the Northweſt windes.*

But when ſhe knew her faithfull fellow ſlaine,
O how ſhe ſhrikt and bruz'd her guiltleſſe arme,
Tearing her haire, renting her cheekes in vaine,
On outward parts, reuenging inward harmes,
Making of teares and bloud a mingled raine,
Wherwith ſhe *Pyram* drencht, & then thus charmes:
   Speake loue, O ſpeake, how hapned this to thee?
   Part, halfe, yea all of this my ſoule and mee.

Sweete loue, reply, it is thy *Thisbe* deare,
She cries, O heare, ſhe ſpeakes, O anſwere make:
Rowſe vp thy ſprights: thoſe heauie lookers cheere,
At which ſweete name hee ſeemed halfe awake,
And eyes with death oppreſt, againe to cleere.
He eyes her once, and eying leaue doth take,
   Euen as faire *Bellis* 2 winkes but once for all,
   When winters 3 vſher haſtneth ſummers fall.

2 *The white Daiſy.*
3 *Harueſt.*

When

and their Flies.

When afterwards she found her scarfe al rent,
His iu'ory sheath voide eke of rapier gilt:
And hath his hand (quoth she) thy soule hence sent?
And was this bloud by this thy rapier spilt?
Vnhappy I: but I no more lament,
But follow thee euen to the vtmost hilt.
   I was the cause of al thy hurt and crosse,
   Hold, take me eke a partner of thy losse.

Whom onely death could from me take away,
Shal death him take from me against my will?
Not so, his power cannot *Thisbe* staye:
Who euen in death wil follow *Pyram* still,
His blade (yet warme) then to her brest she lays,
And falne thereon thus cri'de with crying shrill:
   Parents vniust which vs deny'd one bed,
   Enuy vs not one toomb when we be dead.

And al you heau'nly hostes allot the same:
And thou O tree, which couerest now but one
(One too too hot, for 1 so imports his name)
But couer shalt two carcasses anone:
Weare signes of bloud from both our harts that came
In mourning weed our mischiefes euer mone.
   She dead: Tree, Sires, & Gods gaue what she praide,
   Black growes the fruit, and they together laide.

1 Pyramus sig-nifieth as much as fiery.

D   Since

Of the Silke wormes

<small>1 *Natal.Com.*
lib.vlt.Mytho.</small>

Since which time eke some other (1) Authors faine,
Their humming soules about these haplesse trees,
To be transported from th'Elysian plaine,
Into the snowy milke-white Butterflyes:
Whose seedes when life and moouing they obtain,
How e're they spare the fruit of Mulberies,
   Leaue yet no leaues vntorne that may be seene,
   Because they onely still continude greene.

Yet that there might remaine some *Pyramis*,
And euerlasting shrine of *Pyrams* loue,
When leaues are gone, and summer waining is,
The little creepers neuer cease to moue,
But day and night (placing in toyle their blisse)
Spinne silke this tree beneath and eke aboue:
   Leauing their ouall (2) bottoms there behind,
   To shewe the state of eu'ry Louers mind.

For as in forme they are not wholly round,
As is the perfit figure of the skie,
So perfit loue in mortals is not found,
Some little warts or wants in all we spie,
Nay eu'n as fine and courfe silke there abound,
The best beneath, the worst rold vp more hie,
   So sometimes lust o're-lieth honest loue,
   Happy the hand that keepes it from aboue.

A

and their Flies.

Againe, as these fine troupes themselues deuoure,
Spinning but silken hharses for their death:
VVhich done, they dye therein, (by Natures power
Transform'd to flies that scarce draw one months
So louers sweet is mingled stil with sower, (breath)
Such happe aboue proceeds or vnderneath,
    That still we make our loue our winding sheete,
    VVhilst more we loue, or hotter then is meete.

Others (1) report, there was and doth remaine   1 *Plin.lib.6.*
A neighbour (2) people to the *Scythian* tall,   *cap.*17.
Twixt *Taurus* mount and *Tabis* fruitful plaine,   2 *Called Seres.*
Most iust of life, of fare and diet, smal,
Louers of peace, haters of strife and gaine,
Graye ey'd, redde cheek't, and amber-headed all,
    Resembling rather Gods then humane race,
    Such grace appeard in words, in deeds, and face.

VVhose righteous life and iustice to requite,
(Whether with wind or raine, no man doth know)
God sent vnto them silke-wormes infinite,
In Aprils wane when buds the mulb'ry flow,
Which here and there in euery corner light,
With sixe white feete and body like to snow:
    Eating each leafe of that renowned tree,
    The matter of these silken webbes we see.
              D 2   These

Of the Silke wormes

These webbs for wares they on their coast exchange:
For alien none must come into the Land,
T'infect their people with religions strange,
And file their temples with polluted hand:
Neither do they to other nations range,
New fashions, rites or manners t'understand:
   Better they haue at home, where euery slaue
   Weares silks as rich as here our Princes braue.

   These be the tales that Poetizers sing,
   Of Silken-worme, and of their seed and meate:
1 VVherof only Sweete, I confesse, and drawn from 1 Helique spring,
the muses drank, Full of delighting change, and learning greate.
as Poets imagine. Yet, yet, my Muse dreames of another thing,
   And listeth not of fictions to entreate.
     Saye then (my Ioye) say then, and shortly reede,
     whē silk was made, & how these silkworms breed.

   Was it think'st thou found out by industry?
   Inspir'd by vision or some Angells word,
2 Melchisedec. When first the name of sacred Maiesty,
   Was giuen from heau'n to 2 *Salems* priest and Lord?
   Did not before tenne thousand Silk-worms lye,
   And hang on euery tree their little cord?
     Yes, but (like *Hebrues* narps on *Babels* plaine)
     Vntoucht and vse-lesse there it hang'd in vaine.
                        Before,

Before, most men liu'd, either naked quite,
Or courtly clad in some beasts skinne or hide:
The best were but in linnen garments dight,
Wherein themselues the greatest men did pride:
Yea afterward in time of greatest light, *Mat.13.*
When chiefe Baptizer preach't in desart wide,
   Where said he, silken robes were to be sought,
   But in kings courts? for whome they first were
                               (wrought.

Though whether worme or flye were formed first,
No man so right can tel as wrong presume:
Yet this I hold. Till all things were accurst,
Nothing was borne it selfe for to consume.
No Caterpillers then which venture durst,
To rauish leaues, or tender buddes to plume:
   For onely life and beauty liu'd in trees,
   Til falling man caus'd them their leaues to leese.

The earthly heards and winged posts of skye,
And eu'ry thing that mou'd on Eden ground,
Fed first on hearbs (as Duke of *Horeb* hie, *1 Moses.*
Author of Natures story most profound,
Sets downe to vs for perfit verity,
(Gaines aide of none but fooles and wittes vnsound) *Gen.2.verse 29*
   When for mans foode trees eke allotted were,
   Which from themselues did fruit or berries beare.

                                Durst

Durst then the finest worme but touch the meate,
Or dish which for his soueraigne was ordain'd?
Durst they figges, nuts, peares, plummes or mulb'ri
Before their lord with treaso foule was stain'd? (ea
No certs no, but when ambitious heate,
Reuok't the blisse which sinnelesse Sire had gain'
Then wormes in common fed with vs, and tore
Our trees, our fruits, yea eu'n our selues therefor

---

1 *Herod.*
Act. 12.
2 *Antiochus E-*
piphanes.
3 *Plato, who di-*
ed eaten of lice,
as Diogenes La-
ertius writeth.

Say Romanes heau'nly humane (1) Orator,
Whose words dropt sweeter then *Hymettus* dewe:
Say (2) *Salems* scourge and *Iudaes* tormentor,
Whose very name doth pomp and glory shewe:
Say 3 thou whose writtes men as diuine adore,
Inspir'd from heau'n with knowledge giuen to fev
What are you now? what liuing were you the
But worms repast, though wise and mighty me

Foule-footed bird, that neuer sleepest well
Nor fully, but on highest pearch do'st breathe:
Whose outward shreeks bewray an inward hell,
Whose glistring plumes are but a painted sheathe:
Whose taile, though it with pride so lofty swel,
Yet hides it not thy blacknesse vnderneath.
Tell me: what hast thou got by climing thus,
But to thy selfe a shame, and losse to vs?

and their Flies.

To vs alone? nay ſtowteſt Okes likewiſe,
Hard-harted willowes by the water ſide,
Sweete Cedar wood which ſome thinke neuer dies,
And 1 Daphnes tree though greene in winters tide,   *1 The Bay.*
Yea ſtone, and ſteele, and things of higheſt prize,
From natures womb that flow in greateſt pride:
   What are they albut meate for wormes and ruſt?
   Two due reuengers of ambitious luſt.

Before thou waſt, were Timber-worms in price,   *2 Called Coſſi,*
And ſold for equal weight of pureſt gold?   *which being fat,*
                                                                          were counted
Fed 3 creeping birds one barke-deuouring lice?   *a moſt daintie*
Were ſi'k-worms from 4 *Serinda* brought and ſold?   *diſh in Rome.*
                                                                          Cæl.Sec.lib. 28.
Deuoured they the leaues of tree moſt 5 wiſe,   *An. lect.*
With fury ſuch as now we do behold?   *3 Titmiſe.*
   Rather beleeue as yet they were not borne,   *4 The firſt and*
   Or onely fed on graſſe, on hearbs, or corne.   *princirall place*
                                whence they
                                were brought
                                into Europe.
                                Polyd.virg. lib.
                                11.de inuent.
                                &c.

For ſith their chiefeſt vſe is to arraye   *5 The Mulbery*
This little breathing duſt when time requires,   *is called the wi-*
With gallant guards and broydred garments gaye,   *ſeſt tree, becauſe*
With ſcarfs, vales, hoodes, and other ſoft attires: *it neuer buddeth*
Whoſe ſenſe from ſenſe is fled ſo farre away?   *till all danger of*
Whoſe mind to beare ſo wrong a thought conſpires,   *cold be gone.*
   As once to deeme theſe Silken-mercers ſent,
   When nakedneſſe was mans chiefe ornament?
                                                                   But

Of the silke wormes

But sith they are, and therefore framed were,
Which first was fram'd? the egge? the worme? or flie?
No doubt the flie, as plainely shall appeare,
To all that haue but an indiff'rent eye, (beare,
<small>1 *Euangelus in*</small> Though twoo 1 great Clarks contrary thoughts did
<small>*Macrobius lib.4.*</small> And sentence gaue, without iust reason why,
<small>*sat.cap.3. &*</small>
<small>*Firmus in Plu-*</small>   That egges were made before the hardie Cocke
<small>*tarch.lib.2. symp*</small>   Beganne to tread, or brooding henne to clocke.
<small>*quest.3.*</small>

Pretend they did, that least and simplest things,
(Which none train'd vp in reasons schoole gainsay)
Of things compounded are the formost springs,
Eu'n as a lumpe of rude and shapelesse clay,
Into the mould a Moulder cunning brings,
And by degrees compels it to obey:
  Forming by art what he in mind fore-thought,
  Out of a masse that iust resembled nought.

So eke though egges seeme things confused quite,
And farre vnlike what afterwards they prooue:
Yet formost place they challenge by their right,
For who e're saw a cock or henne to mooue,
Till first they came from out the yolke and white,
And time, and heate, and place, and sitters loue,
  Had formed out a nature from the same,
  Deseruing wel anothers natures name?

Springs

and their Flies.

Springs not from egges that huge 1 Leuiathan,   1 *The VVhale*
The Tortesse eke, and bloudy Crocodile?
Fish, Lyzards, Snakes, and 2 Skippers African,   2 *Locusts or*
VVhose hurtful armies waste the coasts of Nile?   *grashoppers.*
Nay if with one fitte word the world we scanne,
May it obtaine a fitter name or stile,
    Then that we should a common egge it call,
    VVhich giueth life and forme and stuffe to all?

Nay, did not once that cheerefull brooding sp'rite,
Before the earth receiued forme or place,   *Gen. 1. verse 2.*
Sitte closely like a henne both warme and light,
Vpon the wauing nest of mingled masse,
VVhilst yet nights torches had obtain'd no light
Nor Sunne as yet in circled rounds did passe?
    Yes, yes: the words are so apparant plaine,
    That to deny them, were but labour vaine.

These some do vse with other arguments,
To proue that seede and egges were first in time.
VVrested from quires of sacred Testaments,
And those of heathen wittes the chiefe and prime:
VVhich for authentique held by long descents,
If I gainesay, perhaps may seeme a crime:
    Yet rather would I carry crime and scorne,
    Then falsely thinke, imperfect things first borne.
                E   For

Of the Silke wormes

For reason saith, and sense doth almost sweare,
Natures entire to be created furst:
Bodies t'haue beene before the members were,
The sound before the sicke, the whole, the burst,
That confidence had time when lacked feare,
That blessed state fore-went the state accurst:
  Briefly, al bodyes that begotten beene,
  Were not before created bodies seene.

Now what are seedes and egges of wormes or foule,
But recrements of proexisting things,
The bodies burden voyd of life and soule?
Yea, from themselues corruption onely springs,
Vnlesse by brooders heate (as from the whole)
They changed be to belly, feete, or wings:
  Resembling them now metamorphosed,
  In, by, and from whose essence they were bred.

Diphilus and Senecio, their arguments against Firmus and Euangelus, of whom at large in Macrobius and Pliutarke.

Yea, vsual phrase such dreames confuteth quite,
For neuer man, *this is an egges henne* sayd,
But *this a hennes egge is*, shewing aright,
That egges are things by former natures layde,
Begotte of mingled seede by day or night,
Neither with skinne, nor shell, nor forme arrayd,
  Till long they haue abode in natures nest,
  And wearied womb be with their weight oppreſt

A

and their Flies.

Againe, to thinke that seede was made before,
The substance whence it is ingendered,
(Namely from out much nutrimental store,
Through excesse of humours perfited)
Or else to ghesse it formed was of yore,
Ere pipes were laid through which it should be shed,
    What is it but to dreame of day or night,
    E're darknesse were, or any shew of light?

Sith eke all winged creatures by one day,   *Gen. 1. verse 20*
Are elder then the heards that crawle and creepe,   *& 24.*
Conclude with truth and confidence wee may,
All flies were made ere wormes beganne to peepe,
Both they which all day long at base do play,
And night once come, do nothing else but sleepe,
    And these which onely liue to leaue a seede,
    From whence the neuer-idle spinsters breede.

Silke-flies I meane, which not one breast alone,
But all throughout, on head, wings, sides, and feete,
Besides pure white, else colour carry none,
For creatures pure, a colour thought most meete,
Martial'd the first of all in glorious throne,
Whereon shall sit the Lord and Sauiour sweete,
    Who with tenne thousand Angels all in white,
    Shal one day iudge the world with doom vpright

E 2   No

Of the Silke wormes

No spotte on them, as els on eu'ry flye,
Bycause in them no follies euer grew,
No crimson redde doth for reuengement crye,
No wauering watchet, where al harts be true:
No yellow, where there is no Iealousie:
No labour lost, and therefore voide of blue:
  No peachy marke to signifie disdaine,
  No greene to shew a wanton mind and vaine.

No orenge colour, where there wants despight,
No tawny sadde, where none forsaken be:
No murry, where they couet nought but light,
No mourning black, where al reioyce with glee:
In briefe, within, without, they are al white,
Wearing alone the badge of chastity:
  Bycause they onely keepe themselues to one,
  Who being dead, another chuse they none.

True Turtles mine, begotten with the breath,
Not of a lewd lasciuious mortal *Ioue*:   (death,
Whose lawe was lust, whose life was worse then
Whose incests did defile both wood and groue,
But with the breath of him who vnderneath
Rules *Stigian* king, and heau'nly hosts aboue,
  Asist me if I erre in setting forth
  Your birth dayes story; and surpasing worth.
                                Assoone

and their Flies.

Assoone as light obtain'd a fixed seate,
(which equally was first spread ouer all,
Giuing alike, both glistring, shine, and heate,
To euery place of this inferiour ball)
Two master-lamps appear'd in welkin great,
Th'one king of day, whom Poets *Phœbus* call,
    And th'other *Phœbe*, soueraigne of the night,
    Twinnes at one instant bred and borne of light.

Genesis 1.

Him heau'nly Martiall high, in Pallace plac't,
Built all of cleere and through-shining gold,
With columnes chrysolite most brauely grac't,
And flaming rubies, glorious to behold,
Wearing about his yellow-amber wast,
A sloping belt, with studs twise six times told,
    Wherein were grau'n most artificially,
    Twelue stately 1 Peeres of curious imagery.

1 *The twelue signes in the zodiake.*

About him, as in royall Coach hee sate,
Attended Houre, Day, Minute, Month, and yeare,
Spring, Summer, Haruest, Winter, Morning, Fate,
With Instancie, who then was driuer there,
Whipping his fiery steedes from 2 *Libraes* gate,
Not suffring them to stand still any where,
    Saue once in *Gibeon* when fiue kings were slaine,
    By first-made 3 Champiō with their faithles train.

2 *For it was then first haruest and not springtime, as the vulgar sort do hold.*
3 *Iosua cap. 1 e.*

E 3   His

Of the Silke wormes

His sisters court built al of siluer tri'de,
And Iu'ory charret set with Diamons,
Embost with Orient pearles on either side,
Wheeld al with Saphires, shod with Onyx stones,
Declar'd in what great pompe she first did ride
Amongst the other twinckling Paragons,
  Before her honour suffred an eclipse,
  Through serpents guile, and womans greedy lips.

Her handmaids then were perpetuity,
Constant proceeding, and continuance:
No shew of change or mutability
Could iustly then themselues in her aduance:
Her face was ful and faire continually
Not altering once her shape or countenance, (made,
  Till those lights chang'd for whom al lights were
  And with whose fall the heau'ns began to fade.

<small>1 *Oceanus is the king, & his wife Thetis is counted the Queene of the seas.*
2 *The Lady of the riuers.*</small>

Yet still on her wait (1) *Ocean* and his wife,
Nais (2) the faire, and al the watry crue,
Nights, Riuers, Flouds, Springs, hauing else no strife,
Then who may formost proffer seruice due:
Bloud, choller, phlegme, (the rootes and sappe of life)
Are at her beck, waining or springing new,
  According as from thone celestiall,
  She deignes to shine in measure great or small.
                                 When

and their Flies.

When they were crowned now in royall thrones,
And entred in their firſt and happieſt race,
Amongſt thoſe gliſtring pointed Diamons,
Which cut out times proportion, lotte, and ſpace:
Behold the earth with heauy burden grones,
And praies them both to eie and rue her caſe:
   And with their friendly hands and mending art,
   To haſten that which ready was to part.

For eu'n next morne the *All-creating Sire*   *Gen.1.*
Had ſent abroad, I know not I, what word:
Much like to this, *Let Sea and earth conſpire*   ‡ *So called by*
All winged troupes the world for to afford: *Pyndarus, be-*
Wherewith the aire: euen to the deſart fire,   *cauſe nothing*
Was ſo with great and little flyers ſtor'd,   *liues in it.*
   That none but winged people ſawe the eies,
   Of any ſtar or planet in the skies.

O how it ioyes my hart and ſoule to thinke
Vpon the bleſſed ſtate of that ſame daye?
When at a word, a nodde, yea at a winke,
At once flew out theſe winged gallants gay,
Tide each to each in ſuch a friendly linke,
That eu'n the leaſt did with the greateſt playe:
   The doue with hawks, the chickens with the kite.
   Feareleſſe of wrong, rage, cruelty, or ſpite.

Pert

Pert marlins then no grudge to larkes did beare,
Fierce goshawkes with the Phesants had no warre,
Rau'ns did not then the Eagles talens feare,
Twixt Cuckoes and the Titlings was no iarre,
But coasted one another eu'ry where
In friendly sort, as louers woonted were:
 For loue alone rul'd all in eu'ry kind,
 As though all were of one and selfe same mind.

How safely then did these my Turtle-soules
Disport themselues in *Phœbus* cheerefull shine?
How boldly flew they by the iayes and owles,
Dreadlesse of crooked beakes or fiery eyen?
Nay, who in all the flocks of winged foules
Said once in heart, This pris'oner shal be mine?
 When none as yet made other warre or strife,
 Then such as 1 *Hymen* makes twixt man & wife.

1 A Poeticall God, and supposed instructor of brides and bridegroomes.

But since the fall of parents pufft with pride,
Not onely men were stainde in viciousnesse,
But birdes, and beasts, and wormes, and flies beside,
Declining from their former pe<r>fitnesse,
Did by degrees to imperfections slide,
Tainted with pride, wrath, enuie, and excesse:
 Yea, then the husband of one onely henne,
 Was afterwards contented scarse with tenne.
          Hence

Hence, gowts in cocks., and swelling paines appeare,
Hence, Partridge loynes so feeble we do view,
Hence, sparrow treaders liue out scarce a yeare,
Hence, leprosie the Cuckoes ouergrew:
Breefely, none did in true loue perseuere:
But these white Butterflies and Turtles true,
    Who both in life and death do ne're forsake
      Her, whom they once espoused for their make.

They choose not (like to other birds and beasts)
This yeare one wife, another wife the next,
Their choyse is certaine, and still certaine rests,
With former loues their mindes are not perplext,
Hee yeeldes to her, she yeelds to his requests,
Neither with feare nor ielosie is vext:
    She clippeth him, hee clippeth her againe,
      Equall their ioy, and equall is their paine.

Remember this you fickle hearted Sires,
Whom lust transporteth from your peereles Dames,
To scorch your selues at foule and forraine fires,
Wasting your health and wealth in filthie games,
Learne hence (I say) to bridle badde desires,
Quenching in time your hot and furious flames,
    Let little flies teach great men to be iust,
      And not to yeeld braue mindes a prey to lust.

Of the Silke wormes

When thus they were created the firſt day,
Alike in bigneſſe, feature, forme and age,
Cladde both alike in ſoft and white array,
And ſet vppon this vniuerſall ſtage,
Their ſeuerall parts and feates thereon to play,
Amidſt the reſt of natures equipage: (thought)
Who then ſuppos'd (as ſince ſome fooles haue
That little things were made & ſeru'd for nought.

Diſwitted dolts that huge things wonder at,
And to your coſt coaſt daily ile from ile,
To ſee a Norway whale, or Libian cat,
A Carry-caſtle or a Crocodile,

1 Heraclitus, that euer wept.
2 Democritus that euer laughed at the worlds folly.

If leane Epheſian (1) or (2) th'Abderian fat
Liu'd now, and ſaw your madneſſe but a while,
What ſtreaming flouds would guſh out of theyr
To ſee great wittols little things deſpiſe? (eies,

When looke, as coſtlieſt ſpice is in ſmall bagges,
And little ſprings do ſend foorth cleereſt flouds,

3 Called Onis in Engliſh.

And ſweeteſt (3) *Iris* beareth ſhorteſt flagges,
And weakeſt *Oſiers* bind vp mighty woods,
And greateſt hearts make euer ſmalleſt bragges,
And little caskets hold our richeſt goods:
   So both in Art and Nature tis moſt cleere,
   That greateſt worths in ſmalleſt things appeare.
                                        What

and their Flies.

What wife man euer did so much admire
Neroes (1) Colossus fiue score cubits hie,   *1 Made by Ze-*
As *Theodorus* Image cast with fire,   *nodorus : of*
Holding his file in right hand hansomly,   *which, and also*
In left his paire of compasses and squire,   *of Theodorus*
With horses, Coach, and footmen running by   *image, more in Plin.lib.,4.cap.*
   So liuely made, that one might see them all?   *7.& 8.*
   Yet was the whole worke than a flie more small.

Nay, for to speake of things more late and rife,
Who will not more admire those famous Fleas,
Made so by art, that art imparted life,
Making them skippe, and on mens hands to seaze,
And let out bloud with taper-poynted knife,
Which from a secret sheathe ranne out with ease:
   The those great coches which theselues did driue,   *2 Made by*
   With bended scrues, like things that were aliue?   *Gawen Smith. Anno,1586.*

Ingenious (3) Germane, how didst thou conuey   *3 Ioannes Re-*
Thy Springs, thy Scrues, thy rowells, and thy flie?   *gi montanus: of*
Thy cogs, thy wardes, thy laths, how didst thou lay?   *whom Ramus at large in Proem,*
How did thy hand each peece to other tie?   *lib 2.Math.*
O that this age enioy'd thee but one day,
To shew thy Fleas to faithlesse gazers eye!
   That great admirers might both say and see,
   In smallest things that greatest wonders bee.

            F 2   Great

Of the Silke wormes

        Great was that proud and feared Philistine,
        Whose launces shaft was like a weauers beame,
        VVhose helmet, target, bootes, and brigandine,
1 For they weied VVeare weight (1)sufficient for a sturdy teame,
6000 Shekles of VVhose frowning lookes and hart-dismaying eyne,
brasse.   Daunted the tallest king of *Israels* realme:
          Yet little shepheard with a pibble stone,
        Confounded soone that huge and mighty one.

        Huge fiery Dragons, Lions fierce and strong
        Did they such feare on cruel (2)Tyrant bring,
2 Pharaoh. VVith bloudy teeth or tailes and talens long,
        VVith gaping Iawes or double forked sting,
        As when the smallest creepers ganne to throng,
        And seize on euery quicke and liuing thing?
          No, no. The Egyptians neuer (3)feared mice,
3 Yet for feare As then they feared little crawling lice.
of them they
honoured their
Gods in the
forme of cats.
Plaut. lib. de Is.
& osi.
4 A most famous Did euer (4)Pisceus sound his trumpet shrill
trumpeter. So long and cleere, as doth the summer Gnat,
Plin. lib. cap. 56. Her little cornet which our eares doth fill,
        Awaking eu'n the drowziest drone thereat?
5 Anacreon in Did euer thing do *Cupid* so much ill,
one of his latter As once a(5)Bee which on his hand did squat?
Odes.   Confesse we then in small things vertue most,
        Gayning in worth what they in greatnesse lost.
                                       But

But hollà, Muse, extol not so the vale,
That it contemne great hilles, and greater skie,
Thinke that in goodnesse nothing can be small,
For smalnesse is but an infirmitie,
Natures defect, and offspring of some fall,
The scorne of men, and badge of infamy?
   For still had men continued tall and great,
   If they in goodnesse still had kept their seate.

A little dismall fire whole townes hath burnd,
A little winde doth spread that dismall fire,
A little stone a carte hath ouerturnde,
A little weede hath learned to aspire,
The little Ants (in scorne so often spurnd)
Haue galles: and flies haue seates of fixed ire.
   Small Indian gnattes haue sharpe and cruel stings,
   Which good to none, but hurt to many brings.

And truely for my part I list not prayse
These silke-worme-parents for their little sise,
But for those louely great resplendant rayes,
Which from their woorks and worthie actions rise,
Each deede deseruing well a Crowne of bayes,
Yea, to be grauen in wood that neuer dies:
   For let vs now recount their actions all,
   And truth wil proue their vertues are not small.

Of the silke wormes

First, though fiue Males be brought to Females ten,
Yet of them al they neuer chuse but fiue,
Each takes and treads his first embraced henne,
With her he keepes, and neuer parts aliue:
And when he is enclos'd in Stygian penne,
Desireth she one moment to suruiue?
   No, no, but strait (like a most louing bride)
   Flies, lies, and dies, hard by her husbands side.

Anno. Dom. 1579. when I was in Italy.

In Tuscane towres what armies did I view
One haruest, of these faithful husbands dead?
Bleede, O my heart, whilst I record anew,
How wiues lay by them, beating, now their head,
Sometimes their feet, and wings, & breast most true,
Striuing no lesse to be deliuered,
   Then *Thisbe* did from vndesi ed life,
   When she beheld her *Pyram* slaine with knife.

But whilst they liue, what is their chiefest worke?
To spinne as spiders do a fruitlesse threed?
Or Adder-like in hollow caues to lurke,
Till they haue got a curst and cankred seed? (fork,
(Whose yong ones therfore, with dame Natures
Iustly gnaw out the wombs that did them breed:)
   Or striue they Lion-like to seize and pray,
   On neighbours herds or herds-men by the way?
                                   Delight

Delight they with strange 1 Ants & Griphins strong,   1 *Of whō Pliny writeth, lib.* 11.*cap.*31.
To hoord vp gold and eu'ry gaineful thing?
Liue they not beasts, and birds, and men among,
Committing nought that may them damage bring?
O had I that fiue-thousand-versed song,
Which(2)Poet prowd did once with glory sing,   2 *Thamyris, who wrote*
That whilst I write of these same creatures blest,   5000. *verses of*
In proper words their worth might be exprest.   *the worlds creation Zetzes,*7. *childhistor.*108

What wil you more? they feede on nought but aire,
As doth that famous bird of Paradice,
They liue not long, lest goodnesse should empaire,
Or rather through that(3) Hagges enuious eyes,   3 *Atropos.*
That sits, and sitting, cuts in fatall chaire
That threed first off, which faitest doth arise:
    Affording crowes and kites a longer line,
    Then fliers ful of gifts and grace diuine.

When maker said to eu'ry bodied soule,   *Gen.*1.
Encrease, encrease, and multiply your kinde:
What he or she of al the winged foule
So much fulfill'd their soueraigne-Makers minde,
As these two flies? who coupled three dayes whole,
Left on the fourth more seeds or egges behind
    Then any bird: yea then the fruiteful wrenne,   4 *Sometimes, more, seldoms*
    Numbred by tale a(4)hundred more then tenne.   *fewer.*
                                   Which

Of the silke wormes

Which donne, both die, and die with cheerefull hart
Bycause they had done al they bidden were,
Might we from hence with conscience like depart,
How deare were death? how sweet & voyd of feare?
How little should we at his arrowes start?
If we in hands a quittance such could beare
   Before that iudge, who looks for better deedes,
   From men then flies, that spring of baser seeds.

Ψυχὴ is all one name in Greeke for a soule and a butterflie.

Go worthy soules (so (1) witty *Greeks* you name)
Possesse for aye the faire *Elisian* greene:
Sport there your selues ech Lording with his Dame,
Enioy the blisse by sinners neuer seene:
You liu'd in honour, and stil liue in fame,
More happy there, then here is many a Queene:
   As for your seeds committed to my charge,
   Take you no care : I'le sing their worth at large.

2 The Lady of the plaine.
3 Miraes daughter.
4.5 6.7.8. Gentlewomen attending vpon Mira and her daughter.

Weepe not faire (2) *Mira* for this funeral.
Weepe not (3) *Panclea*, *Miraes* chiefe delight,
Weepe not (4) *Phileta*, nor (5) *Erato* tall:
Weepe not (6) *Euphemia*, nor (7) *Felicia* white:
Weepe not sweete (8) *Fausta*: I assure you all,
Your cattels parents are not dead outright:
   Keepe warme their egges and you shall see anone,
   From eithers loynes a hundred rise for one.

              FINIS.

The second booke of the Silke-Wormes and their Flies.

O Thou whose sweet & heau'nly-tuned Psalmes
  The heau'ns theselues are scarce inough to praise!
Whose penne diuine and consecrated palmes,
From wronging verse did *Royall Singer* raise,
Vouchsafe from brothers ghost no niggards almes,
Now to enrich my high aspiring layes,
  Striuing to ghesse, or rather truely reede,
  What shall become of all this little breede.

This little breede? nay euen the least of all,
The least? nay greater then the greatest are:
For though in shew their substance be but small,
Yet with their worth what great ones may compare?
What egges as these, are so much sphericall
Of all that euer winged Natures bare?   *A comparison of*
  As though they onely had deseru'd to haue,   *the Silke flies*
  The selfe same forme which God to heauens gaue. *egges with other egges.*

From *Lybian* egges a mightie (1) bird doth rise,   1 *The Ostrich.*
Scorning both horse and horsemen in the chace,
With Roe-bucks feete, throwing in furious wise,
Dust, grauell, sand and stones at hunters face,
Yet dwels there not beneath the vauted skies,
A greater foole of all the feathred race:
  For if a little bush his head doth hide,
  He thinkes his body cannot be espide.

                  G   From

Of the Silke wormes

1 The Eagle.

From egges of (1) her whose mate supporteth *Ioue*,
And dares giue combate vnto draggons great,
With whom in vain huge stagges and Lions stroue,
Whose onely sight makes euery bird to sweate,
Whom *Romanes* fed in *Capitole* aboue,
And plac't her Ensigne in the highest seate,
   What else springs out but bloudy birds of praye,
   Sleeping al night, and murdering al the daye?

From egges of famous *Palamedian* foules,
And them that hallow *Diomedes* toomb,
In bodies strange retaining former soules,
VVise, wary, warlike, saging things to come,
VVhose inborne skil our want of witte controules,
Whose timely fore-sight mates our heedlesse doom,
   Comes ought but cranes of most vnseemly shape,
   And diuing Cootes which muddy chanels scrape?

2 Peacocks.

Yea (2) you whose egges *Hortentius* sometimes sold,
At higher rate then now we prize your sire:
Proud though he be, and spotted al with gold,
Stretching abroad his spangled braue attire,
VVherby, as in a glasse, you do behold,
His courting loue, and longing to aspire:
   VVhat bring ye forth but spectacles of pride,
   VVhose pitchy feete marres al the rest beside?
                            Thrise

and their Flies.   43

Thrife bleſſed egges of (1) that renowned dame,   1 *The Pelicane.*
Who bleeds to death, her dead ones to reuiue,
Whome enuious creepers poyſon ouercame,
Whilſt ſhe fetcht meate to keepe them ſtil aliue,
How wel befits her loue that ſacred *Lamb*,
That heal'd vs all with bleeding iſſues fiue?
    Yet hath your fruit this blotte, to ouer-eate,
    And glutton-like to vomit vp their meate.

VVinters (2) *Orpheus* bloudy breaſted (3) Queen,   2 *Robin-red-*
Sommers ſweete ſolace, nights (4) *Amphion* braue,   *breſt.*
Linus (5) delight, *Canaries* clad in greene,   3 *VVrenne.*
All (6) linguiſts eke that beg what hart would craue,   4 *Nightin-*
Selling your tongues for euery trifle ſeene,   *gale.*
As almonds, nuttes, or what you elſe would haue: 5 *Linnet.*
    Offſprings of egges, what are you but a voice?   6 *Pies, parratt,*
    Angring ſometimes your friends with too much   *ſtares, &c.*
                                                 (noyſe.

Victorious (7) *Monarch*, ſcorning partners all,   7 *The houſe-*
Stowt lions terrour, loue of martial Sire,   *cocke.*
True farmers clocke, nights watchman, ſeruants call,
Preſſing ſtil forward, hating to retire,
Conſtant in fight, impatient of thral,
Bearing in a little breaſt a mighty fire:
    Oh that thou wert as faithful to thy wife,
    As thou art free of courage voice and life!
              G 2   Chaſte

Of the Silke wormes

Chaste is the Turtle, but yet giuen to hate,
Storkes are officious, yet not voide of guiles,
Hardy are *Haggesses*, but yet giuen to prate,
Faithful are *Doues*, yet angry otherwhiles,
The whitest swimmer nature e're begate,
Suspition blacke and iealousie defiles:
  Briefely, from egges of euery creature good,
  Sprang nought distainted but this little broode.

1 Called by Alchimists Ouum Philosophorum, the Philosophers egge.
2 A medicine famous in Homer to extinguish all kinde of griefes and paines.

As for that (1) egge conceiu'd in idle braine,
Whence flowes (forsooth) that endlesse seed of gold,
The wombe of wealth, the (2) *Nepenthes* of paine,
The horne of health, and what we dearest hold:
I count it but a tale and fable vaine,
By some olde wife, or cousning friar told:
  Supposed true, though time and truth descries,
  That all such workes are but the workes of lies.

For when the Sire of truth hath truly saide,
That none can make the couering of his head,
These slender haires, so vile, so soone decaide,
Of so smal worth though nere so finely spread
Shal any witte by humane art and aide,
Transforme base mettals to that essence redde,
  Which buies, not only pearles and precious stones,
  But kingdōs, states, & *Monarchs* frō their thrones:

Ah

and their Flies.

Ah! heau'ns forbid (nay heau'ns forbid it sure,)
That euer Art should more then Nature breede,
Curse we his worke whose fingers most impure,
Durst but to dare the drawing of that seede,
Yet when they haue done al they can procure,
And giuen their leaden God a golden weede:
   Zeuxis his painted dogge shal barke and whine,
   When *Ioue* they turne to *Sol* or *Luna* fine.

Sisyphian(1) soules, bewitched multipliers,
Surcease to pitch this neuer pitched stone,
Vaunt not of Natures nest, nor *Orcus* fires,
Hoping to hatch your addle egge thereon:
Restraine in time such ouer-prowd desires,
Let cre'tures leaue *Creators* works alone:
   Melt not the golden Sulphur of your hart,
   In following stil this fond and fruitlesse art.

1 Sisyphus was one of king Æolus sonnes, delighted in robbing and cousening of his neighbours, wherefore this punishment was enioyned him, to rowle a stone continually to the top of a Pyramidall and most steepe hil, til it rested there, which was an impossible thing to performe, because he could neuer pitch it. Ouid 3.met.

Record what once befel great *Aeols*(2) sonne,
For counterfetting onely but the sound,
Of heau'nly Canoniers dreadful gunne,
That shakes the beams and pillers of this round:
A fiery boult from wrathfull hand did runne,
Driuing false forger vnder lowest ground:
   Where stil he liues stil wishing to be dead,
   Spotted without, within al staind with redde.

2 Salmoneus, another sonne to Æolus, who for counterfeyting thunder, was turned (as Seruius conceiued) into a Salmon.

Of the Silke wormes

Remember eke the Vulture gnawing still,
That euer-dying euer-liuing (1) wretch,
VVho stealingly with an ambitious will,
From *Phœbus* wheeles would vitall fire reach,
Thinking to make by humane art and skill,
His man of clay a liuing breath to fetch:
  Beware in time of like celestiall rods,
  And feare to touch the onely worke of gods.

1 Prometheus, sonne of Asia & Iaphet, who enterprising (as Paracelsus doth) to make man, was tied vppon mount Caucasus in chaines, there to be eaten euerlastingly by Vultures, and yet neuer to die. Ouid 10. Metam.

But if you still with prowd presumptuous legges,
VVill needes clime vppe the fiery-spotted hil,
Pilfring from *Ioue* his Nectar voyde of dregs,
And that immortall meate (2) which none doth fill,
If ye will needes imbesill those faire egges,
VVhich in her child-bedde did their (3) mother kil,
  Yet say not, that for gifts and vertues rare,
  They do, or may, with these my egges compare.

2 Called Ambrosia.
3 Leda, who being gotten with childe by Iupiter in the forme of a swanne, brought forth two egges, out of the one came Castor and Clytemnestra, out of the other Pollux and Helena. Hesiodus.

These, these, are they, in dream which Romane spide
Clos'd in a slender shell of brittle mould,
Holding within, a white like siluer tride,
VVhose inward yolke resembleth (5) Ophirs gold,
From out whose centre sprang the cheefest pride,
That e're *Latinus*, or his race did hold,
  Exchanging in al countries for the same, (name.
  Meate, drinke, cloth, coyne, or what you else can

4. Cic. 2. de diu.
5 VVhence Salomon fetcht gold euerie three yeares, which wisedome would neuer haue permitted him to haue done, if he had knowne (as some imagine) how to make the Philosophers stone.

Here

and their Flies.

Here lies the (1)Calx of that renowned shel,
Here flotes that water permanent and cleere,
Here doth the oile of Philosophers dwell,
Stil'd from the golden Fleece that hath no peere:
In midst of whose vnseene and secret cell
Dame Nature sittes, and euery part doth steere,
   Though neither opening shop to euery eie,
   Nor telling (2) *Cæsar* she can multiply.

1 Of which Calx, water, and oyle, you may reade more than enough in Libauius: Epist. de ouo Philosophorum, & the troubling Turba Philosophorum, & the reuerent, D Dee, in Monad. Hierogl.
2 As one or two sols haue done.

Al-working mother, Foundresse of this All,
Ten-hundred thousand-thousand-breasted nurse,
Dedalian mouldresse both of great and small,
As large in wealth, as liberall of purse,
Still great with childe, still letting children fall,
Good to the good, not ill vnto the worse,
   VVhat made thee shew thy multiplying pride,
   More in these egges, then all the egges beside?

3 A description of Nature.

VVas it, because thou takest most delight,
To print the greatest worth in smallest things?
That they, the least of any seede in sight,
Might clothiers breed to clothe our mightiest kings?
O witte diuine, O admirable spright!
VVorthie the songs of him that sweetest sings:
   Let it suffice that I adore thy name,
   VVhose works I see, and know not yet the same.
                                 But

Of the Silke wormes

        But damsels, ah : who rustleth in the skie?
1 Boreas, who by   Me thinks I heare *Enitheas* Ladyes (1) foe,
force ranished   Blustring in fury from the mountaines hie,
Orythyia King   Looke how he raiseth cloudes from dust below,
Erictheus
daughter. Ouid   Harke how for feare the trees do cracke and crie,
6. Metam.   Each bud recoyles, the seas turne too and fro:
          O suffer not his breath-bereauing breath,
          To slay your hopes with ouer-timely death.

        Therefore assoone as them you gathered haue,
        Vpon the whitest papers you can find,
        In Boxes cleane your egges full closely saue,
        From chilling blast, of deadly nipping winde,
2 Hyems or   Let not that hoary (2) iry-manteld slaue
winter.   So much preuaile, to kill both stocke and kinde:
          Farre be it from a tender Damsels heart,
          On tendrest seedes to shew so hard a part.

The seedes or   Yet keepe them not in roomes too hot and close,
egges of Silke-   Lest heate by stealth encroch it selfe too soone,
flies are to bee   And inward matter ripening so dispose,
kept neither too
cold, nor any   That spinsters creepe ere winters course be done,
thing hot.   Whilst woods stand bare, & naked ech thing grows,
        And *Thisbes* sap for aide be inward runne:
          For as with cold their brooding powre is spilde,
          So are they then for want of herbage kilde.

                                     Th'Arch-

and their Flies.

Th'Arch-mason of this round and glorious bal,
Of creatures created Man the last,
Not that he thought him therefore worst of all,
(For in his soule part of himselfe he cast)
But left his wisedome might in question fall,
For hauing in his house a stranger plac't,
   Ere eu'ry thing was made to pleafe and fealt,
   So great a Monarch and so braue a guest.

Vnder whose feete where e're he went abrode
Vesta(1) spread forth a carpet voide of art,   **1 The Earth.**
Softer then silke, greener then th'*Emerode*,
Wrought al with flowres, and eu'ry hearb apart,
Ouer him hang'd where e're he made abode,
An azur'd cloth of state, which ouerthwart
   Was biast (as it were) and richly purld,
   With twelue braue signes & glistring stars inurld

Vppon him then as vassals eu'ry day
Stowt Lions waited, tameles Panthers eke,
Fierce Eagles, and the wildest birds of pray,
Huge whales in Seas that mighty carricks wreake,
Serpents and toades: Yea each thing did obey,
Fearing his lawes and statutes once to breake:
   Yet wherto seru'd this pompe and honour great,
   If man had wanted due and dayly meate?

Trace

Of the Silke wormes

The seedes or egs of Silkeflies are not to be hatched till the Mulberie tree be budded.

Trace you Gods steppes, and til you can attaine
Wherwith to feed your guests when first they shew,
Haste not their hatching, for t'wil prooue a paine,
Filling your hearts with ruth, your eyes with dew,
As when th'vntimely lambe on *Sarums* plaine,
Fallne too too soone from winter-starued ewe,
To pine you see for want of liquid food,
Which should restore his wants of vitall blood.

1 The Mullery

Attend therefore, when farmers (1) ioy renues
Her liuely face, and buddeth all in greene,
For Hyems then, with all his frozen crues,
Is fully dead, or fled to earths vnseene.
Corne, cattell, flowers, feare then no heauie newes,
From Northern coasts, or *Boreas* region keene:
   Birds sing, flies buzze, bees hum, yea al things
   To see the very blush of *Morus* lippe.   (skip

2 The Nightingale.
3 The Wrenne.
4 Larkes.

Let swallowes come, let storkes be seene in skie,
Let (2) *Philomela* sing, let (3) *Progne* chide,
Let (4) *Tiry-tiry-leerers* vpward flie,
Let constant *Cuckoes* cooke on euery side,
Let mountaine mice abroad in ouert lie,
Let euery tree thrust foorth her budding pride,
   Yet none can truely warrant winters flight,
   Till she be seene with gemmes and iewels dight.

O

and their Flies.

O peereleſſe tree, whoſe wiſedome is far more
Then any elſe that ſprings from natures-wombe:
For though *Pomonaes* (1) daughters budde before,
And forward (2) *Phillis* formoſt euer come,
And *Perſian* (3) fruit yeeldes of her bloſſoms ſtore,
And (4) *Taurus* hotte ſucceedeth (5) *Aries* roome:
　Yet all confeſſe the Mulbery moſt wiſe,
　That neuer breedes till winter wholly dies.

1 All kinde of round fruit.
2 The Almond tree.
3 Peaches: brought firſt out of Perſia, as Columella writeth.
4 Aprils ſigne.
5 March his ſigne.

Such is her wit: but more her inward might,
For budded newe when *Phœbus* firſt appeares,
She is full leaued e're it grow to night:
With wondrous crackling filling both our eares,
As though one leafe did with another fight,
Striuing who firſt ſhall ſee the heau'nly ſpheares,
　Euen as a liuely chickin breakes the ſhell,
　Or bleſſed Soules do ſcudde and flie from hell.

Yet witte and ſtrength her pittie doth exceede,
For none ſhe hurts that neere or vnder grow,
No not the brire, or any little weede,
That vpward ſhootes, or groueling creepes below,
Nay more, from heauenly flames each tree is freed
That nigh her dwels, when fearful lightnings glow:
　For vertue which, the Romanes made a law,
　To puniſh them that ſhould her cut or ſaw.

So writeth Pliny, lib. 10, hiſt. nat.

H 2　　　I

Of the Silke wormes

Reade Pliny.
lib. citato.

I leaue to tell how she doth poison cure,
From adders goare or gall of Lisards got,
VVhat burning blaines she heales and sores impure,
In palat, Iawes, and al enflamed throte,
VVhat canckars hard, and wolfes be at her lure,
What Gangrenes stoop that make our toes to rotte:
   Briefly, few griefes from Panders boxe out-flew,
   But here they finde a medcine, old or new.

Her bloud retourn'd to sweete *Thisbean* wine,
Strengthneth the lungs and stomacke ouer-weake,
Her clustred grapes do proue a dish most fine,
VVhose kernels soft do stones in sunder breake:
Her leaues too that conuerted are in time,
Which kings themselues in highest prize do reake:
   Thus giues she meat, and drink, medcine, & cloth,
   To eu'ry one that is not drownd in sloth.

1 *So Monardes*
calleth it. lib. de
arb. Ind.
2 *Leo Afer.*

Bragge now no more perle-breeding *Taprobane*,
Of *Cocos* thine, that (1) all-supplying foode,
Vaunt not of Dates thou famous (2) *Africane*,
Though sweete in taste, and swift in making bloud,
Blush *Syrian* grapes, and plums *Armenian*,
Ebusian figges, and fruit of *Phillis* good:
   Bad is your best compared with this tree,
   That most delights my little flocke and mee.

But

and their Flies.

But wil you know, why this they onely eate?
Why leaues they onely chufe, the fruite forfake?
Why they refufe al choife and fortes of meate,
And hungers heate with onely one difh flake?
Then lift a while, you wonder-feekers great,
Whilft I an anfwere plaine and eafie make:
   Difdaine you not to fee the mighty ods,
   Twixt vertuous worms and finful humane gods.

I thinke that God and nature thought it meete,   *Why Silke-*
The nobleft wormes on nobleft tree to feede: *wormes eate on-*
And therefore they elfe neuer fet their feete   *ly Mulberie*
On any tree that beareth fruit or feede: *leaues.*
Others diuine, that they themfelues did weete
No other tree could yeelde their filken threede.
   Iudge learned wittes: But fure a caufe there is,
   VVhy they elfe feede vpon no tree but this.

Ne eate they all, as greedy *Rafers* do,
But leaue the berries to their Soueraigne:
Religioufly forbearing once to bloe
Vpon the fruit, that may their Lord maintaine.
Nay, if thefe leaues (though nothing elfe doth growe
In *Eden* rich their nature to fuftaine)
   Had erft bin giuen for other creatures meate,
   They would haue chufde rather to ftarue then eat.

H 3   In

*Why Silke-
wormes feed on-
ly vpon one
meate.*

In that they onely feede vppon one tree,
How iustly do they keepe dame Natures lore?
Who teacheth eu'n the bleare eyde man to see,
That change of meates causeth diseases store:
The gods themselues (if any such there be)
Haue but one(1) meate, one drinke, and neuer more,
   Whereby they liue in health and neuer die,
   For how can one against it selfe replie.

*1 Called Am-
brosia.
2 Called Nectar*

*3 Read Plutark
4. Symposiav.
quæst. 1*

Dualitie of meates was sicknesse spring,
With whom addition meeting by the way,
Begate varietie of euery thing,
Who like a whore in changeable array,
With painted cheekes (as did *Philinus* sing)
And corall lippes, and breasts that naked lay,
   Made vs with vnitie to be at warres,
   And to delight in discords, change, and iarres.

Wherefore assoone as they beginne to creepe,
Like sable-robed Ants, farre smaller tho,
Blacke at the first, like pitch of Syrian deepe,
Yet made in time as white as *Atlas* snow,
Send seruants vp to woods and mountaines steepe,
When Mulb'ry leaues their maiden lippes do shew:
   Feede them therewith (no other foule they craue,
   If morne and eu'n fresh lesage they may haue.)
                                The

and their Flies.

The firſt three weekes the tend'reſt leaues are beſt,
The next, they craue them of a greater ſize,
The laſt, the hardeſt ones they can diſgeſt,
As ſtrength with age increaſing doth ariſe:
After which time all meate they do deteſt, *So that they eat*
Lifting vp heads, and feete, and breaſt to skies, *not in all aboue*
　Begging as t'were of God and man ſome ſhrowde, *nine weekes.*
　Wherein to worke and hang their golden clowde.

But whilſt they feede, let al their foode be drie
And pull'd when *Phœbus* face doth brightly ſhine, *VVhen their*
For raine, miſt, dewe, and ſpittings of the skie, *meate is to bee*
Haue beene ful of the baine of cattle mine: *gathered.*
Stay therfore, ſtay, til dayes-vpholder flie,
Fiue ſtages ful from Eaſterne *Thetis* line: *That is to ſay,*
　Then leaues are free from any poyſned feede, *till the ſunne be*
　Which may infect this white and tender breede. *fiue houres high*

Keepe meaſure too, for though the beſt you get, *In what quan-*
Giue not too much nor little of the ſame, *titie they are to*
Satiety their ſtomacks wil vnwhet, *be dieted.*
Famine againe wil make them leane and lame:
Lend Witte the knife to quarter out their meate,
As neede requires and reaſon maketh clame:
　Leſt belly break, or meagerneſſe enſowe,
　By giuing more or leſſe then was their due.

Ne

Of the silke wormes

Varietie of meates is naught for them.

Ne chāge their food (us some haue thought it meet)
For Mulb'ries though they are of double kind,
The blacker ones are yet to them most sweete,
From out their leaues most pleasing sappe they find,

1 Boreas, the Northwest wind

But whē they faile whilst *Scythiā* krume 1 doth fleete,
(Turne heau'nly hosts, O turne that cruell wind)
   White Mulb'ry leaues, yea tender Elming bud,
   May for a shift be giuen in steede of foode.

Their table is to be kept cleane.

Sweepe eu'ry morn ere they fresh vittailes see,
Their papred boord, whereon they take repast,
With bundled Time, or slippes of Rosemary,
Leaue nought thereon that from their bellies past,
No not th'alf-eaten leaues of *Thisbes* tree,
And when their seates perfumed thus thou hast,
   Remooue them back againe with care and heede,
   To former place wherein they erst did feede.

The sleepe of Silkewormes.

Oft shalt thou see them carelesse of their meate,
Yea ouer-tane with deepe and heauie sleepe,
Like to that strange and Epidemian sweate,
When deadly slumbers did on *Britons* creepe:
Yet feare thou not, it is but natures feate,
Who netheleffe hath of peerelesse spinsters keepe,
   And makes them thus as dead to lie apart,
   That they may wake and feede with better heart.
                              Thrise

Thrise thus they sleep, and thrise they cast their skin, *How oft they*
The latter stil farre whiter then the rest, *change their*
For neuer are they quiet of mind within, *skinnes.*
Til they be cleane of blackneffe difpoffeft,
Whether becaufe they deeme it fhame and finne
To weare the marke of blackifh fiend vnbleft:
   Or that their parents wearing onely white,
   They therefore in that onely would be dight.

As they in body and in greatneffe grow, *How they are to*
Diuide them into tribes and colonies, *be diftributed*
For though at firft one table and no mo *when they grow*
(Smal though it be) a thoufand wormes fuffice, *greater.*
Yet afterwards (as proofe wil truly fhow)
When they proceede vnto a greater fize,
   One takes the roome of tenne, and feemes to craue
   A greater fcope and portion for to haue.

The loft wherein their tables placed be, *VVhat manner*
Muft neither be too full, nor voide of light, *of roome their*
Two windowes are inough, fuperfluous three, *table muft ftand*
Plac't in fuch fort that one regard the light *in.*
Of *Phœbus* fteeds vprifing as we fee:
And from the other when it drawes to night,
   We may behold them tired as it were,
   And limping downe the wefterne *Hemifphere.*
              I   Glafde

Glasde let them be, or linnen-couerd both,
To keepe out fell and blacke (1) *Monopolites*,
The *Myrmedonian* crue, who voide of sloth
Do wholy bend their forces, toile, and wittes
To priuate gaine, and therefore are ful wroth
To see this nation any good besits:
   Working themselues to death both night & day,
   Not for themselues, but others to array.

1 Ants or Emmets.

The greedy imps of her that slue her sonne,
Pandions (2) daughter; bloudy harted Queene:
The winged (3) steedes in *Venus* coach that runne,
Inflam'd with filthy lust and fires vnseene,
Pursue this flocke, and wish them al vndone,
Bycause they come from parents chaste and cleane:
   O therefore keepe the casements close and fast,
   Lest quellers rage your harmelesse cattle wast.

2 Wrennes and Robins.
3 Sparrowes.

If also carelesnesse haue left a rift,
Or chincke vnstopped in thine aged wall:
Where-through a noysome mist, or rayny drift,
Or poysned wind may trouble spinsters small,
Mixe lime and sand, deuise some present shift
How to repel such cruel foe-men al:
   Small is the charge compared with the gaine,
   That shal surmount thy greatest cost and paine.

and their Flies.

If any seeme to haue an amber coate,
And swell therewith as much as skinne can hold,
Wholy to sloth and idlenesse deuote,
Tainting with lothsome gore the common fold,
Of deadly sick'enesse t'is a certaine note,
VVhose cure, sith none haue either writte or tolde,
 VVisedom commands to part the dead and sicke,
 Lest they infect the faultlesse and the quicke.

How the sicke are knowen from the whole, & in what sort to bee vsed.

Colde sometimes kills them, sometimes ouer-heate,
Raine, oyle, salt, old and wet, and musty foode,
The smel of onyons, leckes, garlick, and new wheat,
Shrill sounds of trumpets, drums, or cleauing woode:
Yea some of them are of such weakenesse great,
That whisprings soft of men or falling floud,
 Doth so their harts and senses ouer-wheele,
 That often headlong from the boord they reele.

Outward causes of their sicknesse

Forbeare likewise to touch them more then needes,
Skarre children from them giuen to wantonnesse,
Let not the fruit of these your precious seedes,
Die in their hands through too much carelesnesse:
VVho tosse and roule and tumble them like weedes
From leafe to leafe in busie idlenesse,
 Now squatting them vppon the floore or ground,
 Now squashing out their bellies soft and round.

  I 2   Thus

Of the Silke wormes

Signes of their readinesse to worke.

Thus being kept and fed nine weekes entire,
Surpriz'd with age ere one would thinke them yong,
With what an ardent zeale and hot desire
To recompence thy trauels do they long?
They neither sleepe, nor meate, nor drinke require,
But presse and striue, yea fiercely striue and throng,
   Who first may find some happy bough or broom,
   Whereon to spinne and leaue their amber loome.

They must secure themselues two daies before you set them to worke.

Then virgins then, with vndefiled hand
Seuer the greatest from the smaller crue,
For al alike in age like ready stand,
Now to begin their rich and oual clue,
(Hauing first paid as Nature doth command,
To bellies-farmer that which was his due)
   For nothing must remaine in body pent,
   Which may defile their sacred monument.

For that is the best and safest way to loose none of them.

So being clensde from al that is impure,
Put each within a (1) paper-coffin fine,
Then shal you see what labour they endure,
How farre they passe the weauers craft of line,
What cordage first they make and tackling sure,
To ty thereto their bottom most diuine,
   Rounding themselues ten thousand times & more,
   Yet spinning stil behind and eke before.

                              None

and their Flies.

None cease to worke: yea rather all contend
Both night and day who shall obtaine the prize
Of working much, and with most speede to end,
Whilst rosie (1) *Titan* nine times doth arise
From purple bedde of his most louing (2) friend,
And eke as oft in (3) *Atlas* vally dies)
    Striuing (a strife not easie here to find)
    In working well, who may exceed their kind.

How they work not aboue nine daies.
1 The sunne.
2 Aurora, the morning.
3 The westerne sea.

Yea some (O wofull sight) are often found
Striuing, in worke their fellowes to excell,
Lifelesse in midway of their trauerst round,
Nay those that longest here do work and dwell,
Liue but a while, to end their threed renownd,
For I haue seene, and you may see it well,
    After that once their bottoms are begunne,
    Not one suruiues to see the tenth dayes sunne.

Go gallant youths, and die with gallant cheere,
For other bodyes shortly must you haue,
Of higher sort then you enioyed here,
Of worthier state, and of a shape more braue,
Lie but three weekes within your silken beere,
Till Syrian dogge be drownd in westerne waué,
    And in a moment then mongst flying things,
    Receiue not feete alone, but also wings.

How they are turned into flies when Dogge daies end, or thereabouts.

I 3   Wings

Of the Silke-wormes

A description of the Silke-flies.
1 An exceeding high hil in Asia
2 Venus Paramour, sonne to Cinara, king of Cyprus, by his owne daughter Myrrha.

Wings whiter then the snow of (1) *Taurus* hie,
Feete fairer then (2) *Adonis* euer had,
Heads, bodies, breasts, and necks of Iuory,
With perfit fauour, and like beautie clad,
Which to commend with some varietie,
And shadow as it were with colour sad,
 Two little duskie feathers shall arise
 From forehead white, to grace your Eben eyes.

When the silke is to be winded from the bottom

Then neither shall you see the bottome moue,
Nor any noyse perceiue with quickest eare,
Death rules in all, beneath, in midst, aboue,
Wherefore make haste you damsels voyd of feare,
Shake off delay, as ere you profit loue,
In boxes straite away your bottoms beare,
 Freed from the coffin wherin late they wrought,
 To gaine the golden fleece you so much sought.

In what sort the silke is to bee winded

First pull away the loose and outmost doune,
As huswiues do their ends of knottie towe,
That which lies vpmost is of least renowne,
The finest threed is placed most below:
Threed fitte for kings, vnmeete for euery clowne,
On Natures quill so wound vp rowe by rowe,
 That if thine eye and hand the end can find,
 In water warme thou maist it all vnwind.

Three

and their Flies.

Three sorts there are, distinct by colours three,
The purest like to (1) their resplendant haire,
Who weeping brothers fal from coursers free,
Their teares were turn'd to yellow amber faire.
The second like (2) her whom impatiencie
Made of a spouse a tree most solitary:
    The last more white, made by the weaker sort,
    Not of so great a price, nor like report.

How many sorts of silke there be.
1 Phaetusa & Lampetia Pha-etons sisters. O-uid 2 Metam.
2 Phillis, De-mophoons spouse turned into an Almond tree.

From out al three, but chiefly from the best,
Are made, not onely robes for priests and kings,
But also many cordial medcins blest,
Curing the wounds that sullen *Saturne* brings;
Which being drunk, how quiet is our rest?
How leaps our hart? how inwardly it springs?
    Speake you sad spirits that did lately feele,
    The hart-breake crush of melancholies wheele.

The use of all sorts of silke.

Nay euen the doune which lies aloft confusde,
Makes Leuant stuffe for country yonkers meete,
Though it of court and cittie be refusde,
And is not worne in any ciuill streete,
But tel me yer, how can (3) he be excusde,
Who trampled eu'n the best with mired feete,
    And in a moment marr'd al that with pride,
    For making which, tenne thousand spinsters dide?

3 Diogenes that dogge, who with his dirtie shooes trode downe Platoes silken Quilt (as Laer-tius writeth) in greater pride then Plato euer vsed it.

Now

Of the Silke-wormes

The first made bottoms are best to be reserued for seede.

Now is of these your bottoms you require,
Some to reserue for future race and seede,
Chuse out the eldest, for their forward fire
Makes-inward flye the sooner, spring and breede:
Whereas the latter ones haue least desire,
And lesser might to perfit *Venus* deede:
    For why, their pride is dul, and spirits colde,

1 *The waining Moone.*
    Borne in the quarter last of (1) *Iune* olde.

Wind none of them, which you for breede allot,
In watrie bath, nor else in wine, or lye,
Lest outward moisture innly being got,
Surrounding, drownes the little infant-flye,
And cause both strings and secundine to rotte,
So that before it liues it learnes to dye:
    Or if you haue them drenched so for gaine,
    At sunne or fire to dry them take some paine.

2 *That is to say, white paper for the first writing paper was the inner rinde of a certaine reede or cane, into which Phillira was transformed. Com. Mes. in Mithol.*

Singled, then laye them on a table neate,
Couered al o're with white (2) *Philliraes* skinne,
Stay them againe till *Phœbus* chariot great
In *Oceans* bath hath twelue times washed bin,
And you shal see an admirable feate,
This form'd and yet transformed broode within:
    From which new shapes new bodies do arise,
    And tailes to heads, and worms are turn'd to flies.

Within 12 daies after the bottoms finished, the silke flies are disclosed.

    Whereat

and their Flies.

Whereat to wonder each man may be bold,
When seely worms themselues new fliers made,
Whilst one anothers face they do behold:
Muse how, and when, & where, this forme they had,
How new hornes sprang frō out their foreheads old,
Whence issued wings, which do them ouer-lade: *Silke flies feede on nothing but aire.*
    For they recording what they were of late,
    Dare not yet mount aboue their former state.

As studying thus they stand a day or more,
Offring to feede on nought but onely aire,
Lothing the meate so much desir'd before,
I meane the leaues of *Thisbes* tree most faire:
Disdaining eke to taste of *Naus* store, *A day or a little*
To quench the heate that might their harts impaire: *more after disclosing, they couple togither.*
    At length they know themselues to be aliue,
    And fal to that for which our wantons striue.

Both long, and longing skud to *Venus* forts, *How long they*
To stirre vp seed that euer may remaine, *are coupled togither.*
He runnes to her, and she to him resorts,
Each mutually the other entertaine,
Ioynd with such lincks and glue of natures sports.
That coupled stil they rest a day or twaine:
    Yea oftentimes thrise turnes the welkin round,
    Ere they are seene vnlocked and vnbound.

K   So

When they die after discipling. So hauing left what e're he could impart,
Of spirits, humors, seede, and recrement,
Willing yet further to haue throwne his hart
Into her breast, to whom he all things ment,
He formost dies and yeelds to fatal dart:
Ne liues she long, but strait with sorrow spent,
(Hauing first laide the egges she did conceiue)
Of loue and life she shortly takes her leaue.

Their egges in colour and bignesse, are likest of all things to Millet seede, wherewith Parrachitos are fed. Smal egges they be, in bignesse, colour, shape,
Like to the meate of *Indian* Parrachite,
Lesse farre in view then seed of garden rape,
In number many, yet indefinite:
For when the females womb begins to gape,
And render what the male got ouer night,
 Now more, now fewer seeds dropt from the same,
 As they were short, or longer at their game.

VVhat number of egges they lay. Yet seldome are they than a hundred lesse,
Sometimes two hundred from their loynes do fall,
Round, smooth, hard-shelld, and voide of brittlenes,
Whited alike, and yellow yolked all,
Whose vertues great no man did yet expresse,
1 The water or riuer wheron all the Muses drinke. Much lesse can I whose knowledge is so smal,
 Though sure I am hence may we find a theame,
 Able to drink vp (1) *Aganippes* streame.

and their Flies.

O keepe them then with moſt attentiue heede,   *How the egges*
From *Boreas* blaſt and *Aeols* inſolence,   *are to be preſer-*
From menſtruous blaſts & breathing keep thē freed,   *ued.*
Auoide likewiſe the mil-dewes influence,
Pray heau'nly *Monarch* for to bleſſe your ſeede,
Helping their weakneſſe with his prouidence:
    So may your milk-white ſpinſters worke amaine,
    When *Morus* lippes ſhal bud and bluſh againe.

And (1) thou whoſe trade is beſt and oldeſt too,   *1 An exhorta-*
Steward of all that euer Nature gaue,   *tion to all Far-*
VVithout whoſe help what can our rulers doo,   *mers and Huſ-*
Though gods on earth appareld wondrous braue?   *bandmen to*
Behold thy helping hand faire virgins wooe,   *plant Mulbe-*
Yea nature bids, and reaſon cake doth craue   *ries.*
    Thy cunning, now theſe little worms to nurſe,
    VVhich ſhal in time with gold fill full thy purſe.

In ſteed of fruitles elms and ſallowes gray,
Of brittle Aſh, and poyſon-breathing vgh,
Plant Mulb'ry trees nigh euery path and way,
Shortly from whence more profit ſhal enſue,
Then from (2) th'Heſperian wood, or orchards gay,   *2 Made and*
On euery tree where golden apples grew: *planted by Æg-*
    For what is ſilke but eu'n a Quinteſſence,   *le, Arethuſa, &*
    Made without hands beyond al humane ſenſe?   *Hyperethuſa,*
                K 2   *King Midas*
                         daughter.

Of the Silke wormes

A commendation of this silke, with that which commeth from the Ossereans, as also with that which is made by the Indian wormes.

A quinteſſence? nay wel it may be call'd,
A deathleſſe tincture, ſent vs from the skies,
Whoſe colour ſtands, whoſe gloſſe is ne're appalld,
Whoſe Mulbr'y-ſent and ſauour neuer dies,
Yea when to time all natures elſe be thralld,
And euery thing Fate to corruption ties:
   This onely ſcornes within her liſts to dwell,
   Bettring with age, in colour, gloſſe, and ſmel.

1 Of theſe Oſſerians or Lords of the wood, read Bonfin. lib. 1 Decad 1. Hung. Hiſt.
2 Aureleanus ſurnamed the Liberall, liuing 274. yeares after Chriſt, in whoſe time a pound weight of ſilke was ſold for the like weight in fine gold. Vopiſcus.

So doth not yours (you (1) Lordings of the woode)
Growing like webbs vppon the long-haird graſſe,
Along the (2) *Oſſerian* bancks of *Scithyan* floud,
Which into *Caſpian* wombe doth headlong paſſe.
No, no: Although that ſilke be ſtrong and good
In outward ſhew, and highly prized was,
   When bounteous *Cæſar* ruled citties prime,
   Yet ſoone it fades, and yeelds to rotte in time.

3 Pauſanias bookes.
4 The Dor-beetle.
5 The Spider.
6 The Reede or cane.
7 The hie oakes.

If (3) bookes be true, there is an *Indian* worme,
As bigge as (4) he that robbs the Eagles neſt,
Shap't like (5) *Arachne* that doth tinſels forme,
And nets, and lawnes, and ſhadowes of the beſt,
Fed with (6) her locks, who yeeiding ſtands in ſtorm,
VVhen (7) woods-ſurueyours lye on earth oppreſt)
   From out whoſe belly, broke with ſurfetting,
   VVhole clewes of ſilk ſcarſe half concocted, ſpring.

Yet

Yet that compar'd with this is nought so fine,
Ne ought so sweetely fum'd with daintie sent,
Nor of like durance, nor like powre diuine:
Mirth to restore, when spirits all are spent,
If it be steept in sweet *Pomannaes* (1) wine,
Till colour fade, and substance do relent:
   Nay, nay, no silke must make that (2) Antidote,
   Saue onely which from spinsters mine is got.

1 The *goddesse of apples*.
2 *Called Confectio* Alkermes *a most singular Electuarie against Melancholie, if it be rightly made.*
3 *Io. Fernelius. lib.7 qui est de compos.med.*

Whereof, if thou a pound in weight shalt take
Vnstaind at all (as *Amiens* (3) floure doth write)
And with the iuce of Rose and pippins make
A strong infusion of some day and night,
Adding some graines of muske and Ambres flake,
And seething all to hony-substance right:
   O what a Balme is made to cheere the heart,
   If pearle, and gold, and spices beare a part?

What neede I count how many winders liue,
How many twisters eke, and weauers thriue
Vppon this trade? which foode doth daily giue
To such as else with famine needes must striue:
What multitudes of poore doth it relieue,
That otherwise could scarce be kept aliue?
   Say Spaniard proude, & tel Italian youth,
   Whether I faine, or write the words of truth.

K 3   Not

Not euer were your princes clad so braue,
Not euer were your wiues deckt as they be,
Much lesse was silk then worne of euerie slaue,   1 *Heliogabalus,*
And artists, sprung from base and low degree,   *for so writeth*
That (1) rioter whose belly diggd his graue,   *Lampridius.*
Clothd all in silke, the Romanes first did see:
Before whose time silke wou'n on linnen threed,   *VVhen the*
Was thought braue stuffe for any Princes weed.   *seede of silke-*
  wormes was first
  brought into
  Europe.

But afterwardes, when holy Palmers twaine   *So Polidor vir-*
From out (2) Serinda brought these worms of fame,   *gil writeth out*
And planted Mulb'ry plants on hill and plaine,   *of Procopius,*
Wherewith to fatte and foster vppe the same: *saying that this*
How rich waxt Italy? how braue was Spaine?   *happened 555.*
In Sattin fine, how braggd each man of name?   *yeares after*
Yea, euery clowne, that euen as now, so then,   *Christ, lib. 3*
Habites did scarce discerne the states of men.   *cap.6.de ret. in-*
  uent.
  2 A citie of east
  India.

Vp Britaine blouds, rise hearts of English race,
Why should your clothes be courser then the rest?
Whose feature tall, and high aspiring face,
Aime at great things, and challenge eu'n the best.
Begge countrymen no more in sackcloth base,
Being by me of such a trade possest:
That shall enrich your selues and children more,
Then ere it did Naples or Spaine before.

No

and their Flies.

No man so poore but he may Mulb'ries plant,
No plant so smal but wil a silke-worme feede,
No worme so little (vnlesse care do want)
But from it selfe wil make a clew of threede,
Ech clew weighs down, rather with more then scant,
A penny weight, from out whose hidden seede,
   (After the winged wormes conception)
   A hundred spinsters issue forth of one.

How easie and chargelesse a thing is is to keep silkwormes.

What ouerplus there is in profite by keeping them.

Diuine we hence, or rather reckon right,
What vsury and proffit doth arise,
By keeping well these little creatures white,
Worthy the care of euery nation wise,
That in their owne or publique wealth delight.
And rashly wil not things so rare despise:
   Yea sure, in time they wil such profit bring,
   As shall enrich both people, priest, and king.

Concerning pleasure: who doth not admire,
And in admiring, smiles not in his hart,
To see an egge a worme, a worme a flier,
Hauing first shewd her rare and peerelesse art,
In making that which princes doth attire,
And is the base of euery famous Mart?
   And then to see the flie cast so much seede,
   As doth, or may, an hundred spinsters breede.
                                  Againe

How great pleasure there is in keeping them, both to the eies, eares, nose, and hands.

Of the silke wormes

Againe to view vppon'one birchen shredde,
Some hundred Clewes to hang like clustred peares,
Those greene, these pale, and others somewhat red,
Some like the locks hanging downe *Phœbus* eares:
And then, how Nature when each worme is dead,
To better state in tenne dayes space it reares:
   Who sees all this, and tickleth not in minde?
   To marke the choyse and pleasures in each kinde.

Eye but their egges, (as Grecians terme them well)
And with a penne-knife keene diuide them quite,
Behold their white, their yolke, their skin, and shel,
Distinct in colour, substance, forme, and sight:
And if thy bodies watchmen do not swell,
And cause thee both to leape and laugh outright,
   Thinke God and nature hath that eye denied,
   By which thou shouldst fró brutish beasts be tried.

When they are worms, mark how they color chãge,
From blacke to browne, from browne to sorrel bay,
From bay to dunne, from dunne to duskie strange,
Then to an yron, then to a dapple gray,
And how each morne in habites new they range,
Till at the length they see that happy day,
   When (like their Sires and heau'nly angels blest)
   Of pure and milk-white stoles they are possest.

                            Large

and their Flies.

Lay then thine eare and listen but a while,
Whilst each their foode from leafage fresh receaues,
Trie if thou canst hold in an outward smile,
When both thine eare and phantasie conceaues,
Not worms to feed, but showrings to distil.
In whispring sort vpon the tatling leaues:
    For such a kind of muttring haue I heard, (teard.
    Whilst herbage greene with vnseene teeth they

When afterward with needle pointed tongue,
The Flies haue bor'd a passage through their clewes,
Obserue their gate and steerage al along,
Their salutations, couplings, and *Adieus*:
Heare eke their hurring aud their churring song,
When hot *Priapus* loue and lust renewes,
    And tel me if thou heardst, or e're didst eye,
    Like sport amongst all winged troupes that flye.

Tis likewise sport to heare how man and maide,
Whilst winding, twisting, and in weauing, thay
Now laugh, now chide, now scan what others saide,
Now sing a Carrol, now a louers lay,
Now make the trembling beames to cry for aide,
On clattring treddles whilst they roughly play:
    Resembling in their rising and their falls,
    A musicke strange of new found *Claricalls*.

               L   The

The smel likewise of silken wool that's new,
To heart and head what comfort doth it bring,
Whilst we it wind and tooze from oual clew?
Resembling much in prime of fragrant spring,
When wild-rose buds in greene and pleasant hue,
Perfume the ayre, and vpward sents do fling,
   Well pleasing sents, neither too sowre nor sweete,
   But rightly mixt, and of a temper meete.

As for the hand, looke how a louer wise
Delighteth more to touch *Astarte* slick
Then *Hecuba*, whose eye-browes hide her eies,
Whose wrinckled lippes in kissing seeme to prick,
Vpon whose palmes such warts and hurtells rise,
As may in poulder grate a nutmegge thick:
   So ioy our hands in silke, and seeme ful loth
   To handle ought but silke and silken cloth.

Such are the pleasures, and farre more then these,
Which head, and hart, eies, eares, and nose, and hands,
Take, or may take, in learning at their ease,
The dieting of these my spinning bands,
Whose silken threede shal more then counterpeise,
Paine, cost, and charge, what euer it vs stands,
   So that if gaine or pleasure can perswade,
   Go we, let vs learne the silken-staplers trade.
                                         But

But lift, me thinkes I heare *Amyntas* fayne,
That shepheards skill wil soone be quite vndone,
Behold faire *Phillis* scuddeth from the plaine,
Leauing her flocks at randon for to runne,
Lo *Lidian* clothier breaks his loomes in twaine,
And thousand spinsters burne their woollen spunne:
    Ah! cease your rage, these spinsters hurt you nought
    But wil encrease you more then ere you thought.

Keeping of silke-wormes hindreth not the keeping of sheepe nor Sheepheards.

For carde an ounce of silke with ten of wooll,
How fine, how strog, how strange a yarne doth rise?
Make trial once, and hauing seene at ful,
Your new found stuffe, chaffred at highest prize,
Then blame your idle heads and senses dull,
Trust not conceit, but credite most your eyes:
    Laughing as much, or more, thē ere you mourn'd,
    When feare you see to ioy and vantage turnd.

Laugh now (faire *Mira*) with thy Virgins white,
For why your egges committed to my care,
Are growne so much in bignesse, worth, and sight,
That Kings and Queens to keep them wil not spare,
Yea Queen of Queenes, for vertue, witte, and might,
Perhaps wil hatch them twixt those hillocks rare,
    Where al the *Graces* feede and *Sisters* nine,
    Who euer loue, and grace both thee and thine.

FINIS.

Apparatus Criticus

Commentary

Glossary

Index to the Commentary

Apparatus Criticus

For *sigla*, see list of copies collated, in the *Textual Introduction*. The abbreviation n. indicates that the emendation relates to a sidenote. *Faults escaped* refers to the printer's errata list (sig. A4ᵛ). Substantive emendations are marked with an asterisk and discussed in the Commentary.

The Table

No. 15   fee de] feede
No. 18   *Relocate* How the sicke are discerned. 59. *at* no. 21 *as second item* How the sicke are discerned. *ibid.*

Book I

2.15	The basest] the basest
3.17	seeemed] seemed
3.20	spring] springs
3.24	whose] Whose
	wood] wood.
5.11	the] thy *Faults escaped*
7.1	earelesse] carelesse
7.3	euer] neuer *Faults escaped*
7.14	courser] Coursers *Faults escaped*
8.9	reward.] reward,
8.17n.	cap. 2] cap. 22.
8.18n.	Pamphia] Pamphila
8.20	fitte] sitte
9.19	priuate] Priuie *Faults escaped*

APPARATUS CRITICUS

10.8	eyes,] eyes.
11.19n.	*Balilon*] Q (u); *Babilon* Q (c)
17.3	his] this *Faults escaped*
17.13	layes] laye *Faults escaped*
17.21	came] came,
19.2	hharses] herses *Faults escaped*
20.10	Silken-worme] Silken-worms
20.16	whē] Whē
21.22	(Gaines aide] Gainesaide
21.22n.	29] 29.
22.22	vnderneath.] vnderneath,
23.9	Timber-worms] (2) Timber-worms
23.12n.	*lib.* 11.] *lib.* 3.
24	*Page number lacking*
24.5n.	*symp*] *symp.*
24.8	ro tread] to tread
25.1n.	*Whale*] *Whale.*
25.18	time.] time,
26.24	opprest] opprest.
27.4	Through] thorough *Faults escaped*
27.15	liueto] liue to
27.24	vpright] vpright.
29.2	(which] (Which
29.10	through] thorough *Faults escaped*
29.18	yeare] Yeare
31.1	now] new
31.13	desart fire] desart (1) fire
31.23	kite.] kite,
34.19n.	*Onis*] *Oris*
34.20	*Ofiers*] *Osiers*
35.2n.	*lib.* 54] *lib.* 34
35.15	coches] (2) coches
35.17n.	*lib*] *lib.*
36.9	strong] strong,
36.15n.	*Plaut.*] *Plut.*
36.17n.	*lib. cap.* 56.] *lib.* 7. *cap.* 56.
39.6n.	*creation*] *creation.*
	108] 108.
39.17n.	*Gen.* 1.] *Gen.* 3.
40.9n.	Ψύχη] 1 Ψύχη

APPARATUS CRITICUS 81

Book II

43.9	Orpheus] Orpheus,
44.20	spread] spread,
46.17	dream] (4) dream
47.1n.	reuerent, D Dee. in] reuerent D. Dee in
47.9	Al-working mother] Al-working (3) mother
48.1	damsals] Q (u); damsels] Q (c)
48.2	Enithean] Erycthean — Faults escaped
48.13*	iry-manteld] icy-manteld
49.16	inurld] inurld.
50.1n.	hatchcd] hatched
51.23n.*	lib. 10, hist, nat.] relocate at 51.11 as lib. 16. hist. nat.
54.6	drinke] (2) drinke
54.6n.	Nectar] Nectar.
54.8	replie.] replie?
54.13	Philinus] (3) Philinus
54.13n.	quaest. 1] quaest. 1.
55.12*	of] oft
55.14n	high] high.
56.1	us] as Faults escaped
56.5	1] (1)
56.5n.	wind] wind.
57.24	limpiug] limping
58.13	Pnrsue] Pursue
59.1	I any] If any Faults escaped
59.9n.	sicknesse] sicknesse.
60.18n.	For] 1 For
61.5n.	Aurora.] Aurora,
61.6	dies)] dies:
61.22	waué] waue
62.1n.	Asia] Asia.
62.9n.	bottom] bottom.
63.6*	solitary] solitayr
64.8*	Iune] Luna
64.18n.	Com. Mat.] Com. Nat.
65.6n.	Relocate at 65.10
65.21	fuch] such
	sports.] sports,
66.15	dropt] drop Faults escaped

67.1n.	*Haw,*] *How*
67.5	fot] for
67.21n.	*daughter.*] *daughters.*
68.9n.	*ehese*] *these*
	Decad] *Decad.*
68.11	*Relocate numeral (2) at* 68.15 = (2) *Caesar*
68.22	opprest)] opprest:
69.5	Pomanaes] Pomonaes
70.9n.	*ret.*] *rer.*
70.11	plauted] planted
71.13	delight.] delight,
71.18	hart.] hart,
73.5	distil.] distil,
73.9	afrerwatd] afterward
73.13	aud] and
73.20	fing] sing
75.2n.	*Sheepe nor,*] *Sheepe, nor*

Commentary

In the lemmata, bold print indicates a word italicized in the text.

Dedicatory Verses

2–4. *Arcadias* **heire**, etc. References to Mary Herbert's part in the preparation of the composite *Arcadia* (1593), her translation of Petrarch's *Trionfi della morte*, and of the *Psalms*. The cosmological metaphor implies the georgic's conventional association with the soil.

5. **bowe, or string will break**. Proverbial (Tilley, B561), but probably also an allusion to Pyrocles' excuse for not keeping up the serious pursuit of knowledge: "the mind itself must, like other things, sometimes be unbent, or else it will be either weakened or broken" (*The Countess of Pembroke's Arcadia* [*The Old Arcadia*], ed. J. Robertson [1973] 14).

9–11. Allusions to the *Iliad*, the *Aeneid*, Statius' *Thebaid*, and Daniel's *The Civile Wars* (1595).

Book I

Page 1.2–3. *brothers bowels*. Robert Sidney (whose *Poems*, ed. P. J. Croft [1984], were probably written in 1597–98), was called "the heire of all [Philip Sidney's] matchlesse worth" by John Davies of Hereford ("The Scourge of Folly," in *Works*, ed. A. B. Grosart [1878] 2: 52).

daughters breast. In the preface to a poem commemorating the birth of Philip Sidney's daughter Elizabeth, Scipio Gentili expressed the hope that the Muses would feed her with the honey of eloquence (*Nereus* [1585] sig. A2).

The **Lady of the plaine**. Moffet's habitual name for Mary Herbert, referring to Salisbury Plain. In a dedicatory sonnet to *The Faerie Queene*, Spenser suggested that Sidney's spirit lived on in her.

2.21–22. The basic meaning is "[we wished our bodies to be clad] because we saw our bodies without clothing, and in so great a state of defilement as reft our wits."

3.11. *to shew their beastly fall.* For the contemporary notion that animal skins worn by Adam and Eve were "evidences of God's mercy, but also symbols of death," see A. L. Williams, *The Common Expositor* (Chapel Hill, 1948) 135.

3.13. *Camels.* Not traced in Pliny. *Camlet* was the "name originally applied to some beautiful and costly eastern fabric" associated by Europeans with camel-hair (*O.E.D.*).

3.20. *springs.* Moffet habitually prefers the exact rhyme (cf. p. 37.16), so I have adopted a manuscript emendation in the Clayton copy in the Bodleian library.

3.22. Plutarch, *De Iside et Osiride,* ed. J. Gywn Griffiths (1970) 352D.

4.13. **Linus.** In Puttenham's *Arte of English Poesie,* ed. Willcock and Walker (1936) 6, he is grouped with Orpheus and Amphion as taming, by music, those who "before . . . remained in the woods and mountains, vagrant and dispersed like the wild beasts, lawlesse and naked, or verie ill clad." His only discoverable connection with flax-processing is the Latin word for flax, *linum.*

5.1. **Tuscane** *Prelate.* Vida's account of the story of Saturn, Phillyra, and Venus closes the first of two books *De bombyce.*

5.10. *th'Idalian tree.* Vida, 1.394–95, has Venus seeking "the shelter of the Idalian wood" because she and Cupid alone, of all the gods and mortals, are naked.

5.19. **Phillyraes eies.** Moffet cites Ovid, *Met.* 6.126, but the main classical source is Hyginus, *Fabulae* 138.

6.3–4. Plants commonly found in English herb-gardens, cultivated for their medicinal properties; all are described in Dodoens' *Niewe herball* (1578 trans.).

6.17. *Thus thus perplext.* Possibly corrupt, but cf. p. 13.16.

8.6. *heigth and depth.* For Saturn's institution of the golden age in Latium, see Natalis Comes, *Mythologiae libri x,* 2.2 (Venice, 1581), and Macrobius, *The Saturnalia,* ed. P. V. Davies (New York, 1969) 1.7.21–25.

8.17. **Vespasians** *Scribe.* Pliny, *Natural History* 11.76. The forms *Cean* and *Latous* are derived from the *Castigationes Plinianae* of Ermolai Barbari (Rome, 1492–93; textual commentary to Bk. 11, ch. 22), which were incorporated into the edition by J. N. Victorius (Lyons, 1563) 11.22.200. Modern editions refer to the island of Cos and Plateus.

9.9. **Loves Schoolemaster.** "Praeceptor Amoris" (*Ars Amatoriae* 1.17). The story of Pyramus and Thisbe is told in *Met.* 4.55–166.

COMMENTARY

9.19. *Privie plants.* Probably trees planted as a high hedge to secure privacy. Sir Hugh Platt, *The garden of Eden* (published 1653) 71, recommends, "instead of privy hedges about a quarter," a fence of fruit trees.

11.9. *Rose-fingred Dame. Odyssey* 4.306 and *passim,* not in the *Iliad* 4 as Moffet's sidenote claims.

11.19. **Ninus** *tombe.* M. A. Coccius (Sabellicus), "Enneades," 1.1, in *Opera* 1.14 (Basle, 1560).

11.22. *speedy* **Ister**. Cf. Sidney's OA 66.23, "Languet, the shepherd best swift *Ister* knewe."

11.24. Cf. Ovid, *Met.* 10.446–47: "It was the time when all things are at rest, and between the Bears Bootes had turned his wain with down pointing pole" (Loeb translation by F. J. Miller, who comments, "At midnight these constellations attain their highest point in the heavens, and thereafter begin their downward course").

12.7. **Lethes** *brother.* Possibly Cicero, *De natura deorum* 1.11.28.

12.22. **Lucines** *eye.* For the dramatic irony, cf. Milton's "Epitaph on the Marchioness of Winchester," 28, "Atropos for Lucina came."

13.16. Presumably represents Pyramus' uncertainty, "If this is what I am to think, then . . .; yes, this *is* how I should take it."

14.3. *thrise-accursed.* Ovid gives the three elements more clearly: "I ordered you to come to a place full of danger (1), at night (2), and I was not here to meet you (3)."

15.6. "Blood seeped into the ground and mournfully encircled the mulberry tree's root."

16.7. **Corus**. More precisely Northwest by West. Thomas Cooper's *Thesaurus* (1565) quotes *violentus Corus* from Lucan.

16.23. **Bellis**. The Margarite Daisy, closely associated with spring.

17.13. *laye.* Not recorded by the *O.E.D.* as past tense for "to lay," but I adopt the correction from "Faults escaped."

17.19. *Pyramus* is derived from Greek, πῦρ (fire).

18.3. *th'Elysian plaine.* Natalis Comes, *Mythologiae libri x* (Venice, 1581) 680, describes it as a place where young men and maidens sing together "most beautifully."

18.15–20. Synesius of Cyrene, *Calvitii Encomium* (Praise of Baldness), in *Essays and Hymns,* trans. A. Fitzgerald (1930) 2:253–54, develops the conceit that whatever is perfectly spherical, like the sky, is of divine origin, "Now, what could be balder than a sphere, and what more divine?"

18.22. *best beneath.* See p. 62.17–20.

19.6. Meaning obscure, probably glancing at the notion that a "middle Fate" is best.

19.10. *neighbour people.* By ancient report, the Chinese (Seres) were a wise, peaceful and secluded people. See Pliny, 6.53–54 and 88; Ammianus Marcellinus, 23.6.67–68, adds that silk is worn even by the lowest class of Chinese (some commentators apply this remark to Rome in the second century AD). G. F. Hudson, *Europe & China* (1931) 80–81, conjectures that Pliny's physical description of the Seres, repeated here, may refer to the tribesmen of the Pamirs, who normally regulated the silk trade from Kashgaria to Bactria.

20.9. *Poetizers.* Francis Bacon, *Advancement of Learning,* ed. W. Aldis Wright (1868) 1.4.10, and Thomas Browne, *Religio Medici,* ed. L. C. Martin (Oxford English Texts, 1964) 2.15, treat Pliny as an inventor of fiction.

20.11. Propertius, *Elegies* 3.3.

20.20. **Salems** *priest and Lord.* Gen. 14:18–20.

21.6. *chiefe Baptizer.* Matt. 11:7–8. The words are in fact Christ's.

21.19. *Duke.* Used in the Geneva Bible to render the Vulgate's *dux,* e.g., Gen. 36:15. Moffet's interpretation of Gen. 2:29 to mean that only man could feed on trees is provocatively literalist.

22.9. *Romanes heav'nly human Orator.* Acts 12:22. "*Hymettus* dewe" is honey; cf. Tilley, T391.

22.11. **Salems** *scourge.* Antiochus IV Epiphanes, whose name is glossed as "noble" in the Geneva Bible (1 Macc. 1:11).

22.13. Diogenes Laertius, 3.40.

23.9. *Before thou wast.* Probably means "before the Fall."
Timber-worms. L. Coelius Richerius, *Lectiones antiquae libri tredecim* 28.2.1065 (Basle, 1566) and Pliny, 17.220.

23.12. See p. 70.9–10.

24.5. *twoo great Clarks.* Plutarch, *Quaestiones convivales* (or *Symposiaca problemata*) 2.3, 635E–638A, and Macrobius, *The Saturnalia,* ed. P. V. Davies (New York, 1969) 7.16.1–14, where Disarius debates the question in response to Evangelus' question.

25.1. *Leviathan.* On the contrary, whales were known to be viviparous (Pliny, 9.41).

25.7. *common egge.* The notion of the world as an egg was widespread; see E. M. W. Tillyard, *The Elizabethan World Picture* (1943; reprint 1972) 53, and the *Ovi encomium* of Erycius Puteanus (1574–1646; in C. Dornavius, *Amphitheatrum,* Hanover, 1619).

27.9–10. Gen. 1:20 and 24 are rather ambiguous about the creation of worms. The Geneva Bible's gloss includes them in the fifth day with sea and air creatures and "that which crepeth."

29.8. *Iliad* 4.101.

29.21. **Libraes** *gate.* Moffet's marginal note is reinforced by Benedictus Pererius, *Commentariorum et disputationum in Genesim tomi quatuor* (Cologne, 1622) 38–41, "An mundus verno tempore fuerit conditus."
29.23. Joshua 9–10.
30.21–22. For the moon's influence on bodily humors, see Macrobius, *Saturnalia* 7.16.25–27.
31.13. *desart fire.* Not traced in Pindar (see sidenote), unless Moffet alludes to *Pythian Odes* 1.21, "unapproachable fire."
32.4. *Titlings.* In *Healths Improvement,* 105, Moffet calls the Titling "*unfortunate Nurse* (for the Cuckoe ever lays his egg in the Titlings nest)."
32.11. Silk-moths cannot in fact fly, though this imperfection might be supposed to postdate the Fall.
34.9–37.24. This passage is reprinted by Scoular, *Natural Magic* (1965) 38–41; for the theme, cf. Thomas Browne, *Religio Medici,* ed. L. C. Martin (1964) 1.15:

> Ruder heads stand amazed at those prodigious pieces of nature, Whales, Elephants, Dromidaries, and Camels; these I confesse, are the Colossus and Majestick pieces of her hand; but in these narrow Engines ["Bees, Aunts, and Spiders"] there is more curious Mathematicks, and the civilitie of these little Citizens, more neatly set forth the wisedome of their Maker; Who admires not *RegioMontanus* his Fly beyond his Eagle . . . ?

A "wild cat" is used as a locution for a natural wonder by Lancelot Andrewes, *Sermons,* ed. G. M. Story (Oxford English Texts, 1967) 111, "*Christ,* is no wild catt" (that is why we "love to make no very great haste" to adore Him). The elephant was admired for sleeping on its feet, and the crocodile for moving its upper jaw.
35.2–3. Pliny 34.45–47, 83.
35.15. *coches.* A. S. T. Fisher, "The Source of Shakespeare's Interlude of Pyramus and Thisbe," *N & Q* 194 (1949): 400–401, conjectures that "Gawen Smith" may be the man mentioned twice in the *Private Diary of Dr. John Dee* (1842 ed.) 32, 35.
35.17. *Ingenious Germane.* Regiomontanus, adopted name of Johann Muller of Konigsberg (1436–76), the famed German mathematician. His artificial fly would take off from its maker's hand, fly to each of his dinnerguests, and return, as if exhausted. See Peter Ramus, *Scholarum Mathematicarum libri xxxi* (Basle, 1569) 65–66, and du Bartas (trans. Sylvester) 1.6.885–906.
36.15. For the Egyptians' fear of mice, see Aelian, *Hist. anim.* 6.41, and for

their cat-goddess Bastet, Plutarch, *De Iside et Osiride,* 376C–F, and Herodotus, 2.66–67.

36.17. *Piseus.* Pliny, 7.201.

36.21–22. **Cupid.** For the popularity of the *Anacreontea* 40, see James Hutton, "Cupid and the Bee," *PMLA* 56 (1941): 1036–58.

37.1. *extol not so the vale.* Mantuan's Eighth *Eclogue* begins with an argument between Alphus and Candidus commending hill and vale respectively.

37.9–16. Cf. Dyer's "A Modest Love," in *The Oxford Book of Sixteenth Century Verse.* Scoular cites Tilley, F274, "A little fire burns up a great deal of corn"; W424, "A little wind kindles, much puts out the fire"; S884, "A little stone in the way overturns a great wain"; W238, "An ill weed grows apace"; F393, "The fly has her spleen, and the ant her gall"; and V67, "No viper so little but has its venom."

38.10 (sidenote). Moffet in fact toured Italy in June and July, 1580. See Manfred E. Welti, "Englisch-baslerische Beziehungen zur Zeit der Renaissance in der Medizin, den Naturwissenschaften und der Naturphilosophie," *Gesnerus* 20 (1963): 123–24.

38.18. *fruitlesse threed.* The silk fibre spun by spiders is not commercially exploited. Bacon, *Advancement of Learning,* ed. W. Aldis Wright (1868) 1.4.5, compares the productions of the schoolmen to cobwebs, "admirable for the fineness of thread and work, but of no substance or profit."

39.1. Pliny, 11.111 and 33.66; one of Thomas Browne's "vulgar errors" (*Pseudodoxia Epidemica,* ed. R. Robbins [Oxford English Texts, 1981] 3.11).

39.6. *Poet prowd.* Joannes Tzetzes, *Variarum historiarum chiliades* (Basle, 1546) 7.108; cf. *Iliad* 2.595.

40.9. *soules.* Aristotle, *Hist. anim.* 5.19 (551a); see W. T. M. Forbes, "The Silkworm of Aristotle," *Classical Philology* 25 (1930): 24–25.

40.17–21. **Mira** is an anagram of Mary (Herbert). In attributing pastoral pseudonyms to Mary Herbert and her attendants, Moffet possibly follows the example of Abraham Fraunce in *The Countesse of Pembrokes Yvychurch* (1591) sig. E1^v.

Book II

Page 41.17–44.8. This catalogue of birds is chiefly derived, both for detail and order, from Pliny, 10.

42.17. **Hortentius.** Pliny, 10.45. A Roman orator of Cicero's time, famed for his collection of exotic birds and animals.

43.11. **Linus** *delight.* Another quibble on Linus' name. Cf. p. 4.13.

COMMENTARY 89

43.15. *but a voice.* "*Vox et praeterea nihil*," a proverb in Latin and Greek. See Plutarch, *Apophthegmata Laconica* 233A.
43.18. *lions terrour.* Another of Browne's "vulgar errors"; see the *Pseudodoxia Epidemica,* ed. R. Robbins (1981) 3.27.7. Edward Topsell, *The historie of foure-footed beastes* (1607) 464-65, accounts for the lion's fear by arguing that both cock and lion partake of the sun's "qualities in a high degree," so that the (small) cock's "sunny propertie" is more concentrated than the lion's.
44.5-6. In *Healths Improvement,* pp. 87 and 106, Moffet characterizes the swan as hypocritical because of its white exterior and black flesh.
44.9 (sidenote). **Philosophers egge.** Alchemists likened their process to an egg because all metals reduced to a "tincture" were thought to be gold *in potentia,* just as an egg is a potential creature. See Jonson's *The Alchemist,* ed. F. H. Mares (1967) 2.3.131-36.
44.10. *seed.* The "multiplication" of gold was symbolized by the depositing of the male "seed" of gold in the "womb" of silver. See E. H. Duncan, "Jonson's *Alchemist* and the Literature of Alchemy," *PMLA* 61 (1946): 708-09.
44.11. **Nepenthes.** The philosophers' stone was reputed to cure all diseases. Michael Maier (1568-1622) describes his quest for this universal medicine, which would be able to cure even grief and anger, in "A Subtle Allegory concerning the Secrets of Alchemy" (in *The Hermetic Museum* [1625] trans. A. E. Waite [1893] 2:199-223).
44.22. *essence redde.* Sulphur, which (alchemists believed), mixed with mercury over fire, will produce gold.
45.2. Cornelius Agrippa, *Of the vanitie and uncertaintie of artes and sciences* (trans. J. Sandford, 1569) fol. 157, identifies this as the "vanitie" of alchemical labors. Art, he argues, can only imitate, not surpass, nature.
45.6-8. A play on the metals associated with Saturn, who is "leaden," symbolizing dullness and melancholy, but whose rule is identified with the golden age. Jove (line 8) represents tin. I have traced no reference to "Zeuxis his painted dogge"; instead of birds responding to a life-like painting, here a painting itself responds to outrageous human behaviour.
45.9. Conventionally, Sisyphus was used as a warning to alchemists not to betray the confidentiality of their craft; see *The Alchemist* 2.3.207-10. But Robert Sidney endorses Moffet's satirical intention here in his "endless alchemist" sonnet (xvii): "[I] hope and want, and strive and fail."
45.17-20. An allusion to the alchemists' claim to be able to make artificial thunder, e.g., in *The Alchemist* 1.1.59-62, and Paracelsus, *The Alchemical and Hermetical Works,* trans. A. E. Waite (1894) 1.42n. Servius, in

his commentary to the *Aeneid* (ed. G. Thilo and H. Hagen [Leipzig, 1883]) 4.696; 6.585–94), does not mention the salmon, although the conjectural derivation of "salmon" from *salire*, "to leap," is pertinent. Moffet's description of the salmon may derive from Ausonius, *Mosella* 97–105.

46.1–6. Andreas Libavius, *Rerum chymicarum epistolica forma descriptarum liber primus* (Frankfurt, 1595) 107–16, took Prometheus as the prototype of the alchemist. Paracelsus claimed to be able to produce a "homunculus," a man outside the womb, by freezing semen in a cucurbite (*Works*, trans. Waite, 1.124–25).

46.14 (sidenote). *Leda*. The distribution of progeny in Moffet's note is that given by Natalis Comes, *Mythologiae libri x* (Venice, 1581) 2.1 rather than Hesiod.

46.17. Cicero, *De divinatione* 2.65.

46.20. *Ophirs gold*. Solomon sent his ships to Ophir to fetch gold worth 3.6 m. crowns (see Geneva Bible gloss to I Kings 9:16–18) on a round trip of about three years. It was popularly believed that Solomon had written about the philosophers' stone. See *The Alchemist* 2.1.82.

46.22. Latinus claimed to revive the golden age of his ancestor Saturn. His negotiations with Aeneas in the *Aeneid* 7 are marked by sumptuous hospitality and exchange of gifts.

47.1–3. The allegory of the philosophers' egg attracted the speculation of many alchemical thinkers. Andreas Libavius, *Rerum chymicarum epistolica forma descriptarum liber primus* (1595) 179–84, "De ovo philosophico," repudiated the opinion that salt, mercury and sulphur (the three prime elements) could literally be produced from the calx (calcined egg-shells), the white and the yolk of the egg respectively. But he accepted the symbolic truth of the legend, because the alchemist's apparatus was egg-shaped, and in it the chemical oil (corresponding to the yolk) and water (corresponding to the white) were heated, and a powder, somewhat like the calx, precipitated. Libavius also wrote a treatise on the silkworm, published in his *Singularium pars secunda* (Frankfurt, 1599) 364–524.

In the anonymous compilation *Turba philosophorum* (1572), trans. A. E. Waite (1896) 11–12, the egg was taken to be an allegory of the world, including the four elements, earth (shell), air (membrane between the shell and the white), fire (yolk) and water (white).

John Dee, *The Hieroglyphic Monad* (1564), trans. J. W. Hamilton-Jones (1947) 24–28, ridiculed the alchemists' attempts to decipher the elements of the hieroglyphic philosophers' egg, which he claimed might have an astrological significance.

47.4. *golden Fleece.* From the time of Suidas onward, it was commonly thought that the object of Jason's quest was an alchemical treatise. Vida, 2.376-78, compares silk to the golden fleece.

47.8 (sidenote). **fooles.** Possibly refers to offers made to Queen Elizabeth by Cornelius Lannoy in 1565 and Roloff Peterson of Lubeck in 1594 (*C.S.P. Domestic* [1549-80] 249, and [1591-94] 435). See J. B. Black, *The Reign of Elizabeth* (Oxford History of England, 2nd ed. [1959]) 310-11.

47.9-14. Lucretius, *De rerum natura* 5.783-836, and Alanus de Insulis, *De planctu naturae,* Metre 4, *O Dei proles, genetrixque rerum.*

48.1. *damsels.* The address to the silk-maidens marks the shift to practical instruction, by association with Vida's *De bombyce,* in which the *seriades* are prominent.

48.2. Ovid, *Met.* 6.682-710.

48.13. *icy-manteld.* I have emended in accordance with Moffet's source, du Bartas, who describes Boreas (Orythia's love) as having a "hoarie" beard and in the same passage describes Winter's "white Freeze mantle freng'd with Ice below" (trans. Sylvester, 1.4.694-97). The *c/r* confusion is common in secretary hand.

49.1-8. An imitation of du Bartas, 1.6.427-36.

50.16. **Morus.** Latin for the mulberry tree.

50.19. **Tyry-tiry-leerers.** An *O.E.D.* first citation; but Moffet's use is probably derived from du Bartas, 1.5.615-16.

50.20. *constant* **Cuckoes.** Cf. Tilley, C891, "The cuckoo sings all the year."

51.1-52.24. Scoular, pp. 41-42, reprints the first three stanzas of this section, which is a rhetorical set-piece of formal commendation. But Moffet is not unconscious of the irony of the mulberry's Latin name, *morus* (as in p. 50.16), which can also mean "a fool." See *Healths Improvement,* 208.

51.4. The almond is the first tree to bud in winter (Pliny, 16.103).

51.5. Columella, *De re rustica* 10.409-10.

51.9-16. The description of rapid budding is translated from Pliny, 16.102, to which the sidenote at line 33 presumably refers (hence my emendation of the reference), but the fanciful simile is original.

51.17-24. In this stanza Moffet appropriates to the mulberry two extraordinary powers normally attributed to other trees: the beneficial shade of the elm, and the protection from lightning afforded by the bay-tree and the fig (Pliny, 2.146; 15.134; 17.89-91).

52.1-8. Pliny, 23.134-40. Moffet uses two metaphors from hawking, "at her lure" and "stoop," to highlight the medicinal power of the mulberry.

52.18. **Cocos.** The reference is probably to Garcia ab Orta, *Aromatum, et simplicium aliquot medicamentorum apud Indos nascentium historia* (Antwerp,

1579) 101–10. In the Bodleian Library copy, this work is bound together with Nicolaus Monardes' *Simplicium medicamentorum ex novo orbe delatorum historia* (Antwerp, 1579); both Latin translations are by Charles de L'Ecluse. Orta asserts, "I think there is no tree that can be found more suitable for human use than the Indian nut." Cf. du Bartas (trans. Sylvester) 1.3.841–58.

52.19 (sidenote). **Leo Afer.** John Leo Africanus, or Hasan ibn Muhammad, *A geographical historie of Africa* (trans. John Pory [1600]) 264–80.

54.9–16. Plutarch, *Quaestiones convivales* 4.1.660–64. Moffet's approval here of a simple diet is in marked contrast with his elaborate refutation of the doctrine in *Healths Improvement*, chapter 28.

54.19. *Syrian.* Possibly a color, mentioned by Pliny, 35.30.

55.12. *oft.* One of two consecutive *t*'s ("oft the") has been overlooked. "Oft" makes better sense and bears the metrical stress.

56.5 (sidenote). **Northwest.** Convention dictates that the location of the poem should be Mediterranean.

56.19. *Epidemian sweate.* A reference to the sweating fever, *Anglicus sudor*, which in 1551–52 prompted John Caius (whom Moffet later knew) to write *A boke or counseill against the disease commonly called the sweate* (ed. E. S. Roberts, 1912).

58.10. **Pandions** *daughter.* Procne; hence, the wren. Ovid, *Met.* 6.634.

60.14. *bellies-farmer.* A "dung-farmer" collects and sells manure, so a "bellies-farmer" here must be whoever collects the silkworms' excrement. The worms' purging themselves before beginning to spin contributes to the purity of the final product.

61.17–24. The transformation of the silkworm into a moth was a common emblem of immortality or spiritual rebirth: see Joachim Camerarius, Jr., *Symbolorum centuria tertia* (Nuremberg, 1596) fol. 96; Thomas Browne, *Religio Medici,* ed. L. C. Martin (1964) 1.39; and Henry Vaughan's "Resurrection and Immortality" (*Silex Scintillans*, 1650).

62.21. *unmeete for every clowne.* But cf. pp. 20.7–8; 70.15–24. In *The Theater of Insects* (1658 trans.), Moffet deplores the promiscuous use of silk: "So often as I consider, that some ten thousands of Silk-worms labouring continually night and day, can hardly make three ounces of Silk, so often do I condemn the excessive profusion and luxuriousness of men in such costly things." English sumptuary laws are described by Frances E. Baldwin, *Sumptuary Legislation ... in England* (1926) 433–49.

63.6. *solitayr.* Both the rhyme and the metre require the transposition of *r* and *y*. The ms probably read "solitaire," which the compositor read as "solitarie" and set as "solitary."

COMMENTARY 93

63.21–24. Diogenes Laertius, 6.26.
64.8. **Luna**. *June* is metrically defective and inconsistent with the gloss, "waining Moone." Both Lazarellus, *Bombyx,* pp. 101–2, and Vida, 1.69–81, insist that the silkworms' eggs should be hatched only at full moon.
64.10. *watrie bath*. This process is described by Robert Dalington, *A survey of the great dukes state of Tuscany, in 1596* (1605) 33.
64.18 (sidenote). Not traced in Natalis Comes, *Mythologiae libri x* (Venice, 1581).
65.20. Cf. *Measure for Measure* 1.2.143–44; Moffet's phrase is derived from Vida, 2.302–3:

> *dulcique fruuntur*
> *Amplexu caudis ambo per mutua nexis*
>
> Each enjoys the other mutually in a sweet
> embrace, with bodies entwined.

67.9–24. This appeal to husbandmen helps to locate the poem in the georgic tradition.
67.21. *th'Hesperian wood*. The garden of the Hesperides is associated with the alchemists' "quintessence" in *The Alchemist* 2.1.101–04, and Michael Maier's "Subtle Allegory Concerning the Secrets of Alchemy," in *The Hermetic Museum,* trans. A. E. Waite (1893) 2.208.
68.9–16. Antonius de Bonfinis' description of the "Osseres, Latine sylvarum nobiles" in *Rerum Hungaricum decades quatuor cum dimidia* (Frankfurt, 1581) 15, is a corruption of Pliny's "Seres, lanicio sylvarum nobiles" (see commentary to p. 19.10), and it seems likely that Moffet here deliberately invents, only to dismiss, a rival to the silkworm. The notion that silk was made from trees, by sprinkling the leaves with water and combing the white down of the leaves, was widespread in classical antiquity.
68.15. Vopiscus, *Divus Aurelianus* 45, in *Historia Augusta.*
68.17–24. The passage from Pausanias, *Description of Greece* 6.26.6–9, that provides the details for this stanza is actually an attempt to describe the silkworm itself (see J. G. Frazer's commentary, London, 1930). Again, the subject is the Seres, and again it is likely that Moffet is mischievous in distinguishing this *"Indian* worme" from the silkworm.
69.10. Joannes Fernelius of Amiens, "De compositis medicamentis," *sub* "De antidotis," *Therapeutices universalis, seu medendi rationis, libri septem* 7 (Frankfurt, 1575) 380–81. See also the flattering description of "Alkermes" by Henry Tubbe (d. 1655) in his poem "On the Silke-worme," in *Selected Works,* ed. G. C. Moore Smith (1915).

70.5–8. Lampridius, *Elagabalus* 26.1, in *Historia Augusta*. The sense is, "Elagabalus was the first man the Romans saw clad entirely in silk." Oriental silk was normally rewoven with linen or wool for export to the West. See G. F. Hudson, *Europe & China* (1931) 77–93, esp. 91–92.

70.9–10. The reference here to Polydore Vergil, *De rerum inventoribus* 3.6 (1576) 178, is accurate, in contrast with that at p. 23.12 (which I have emended). Unlike Vida, 2.381–end, who tells the fable of an Eastern prince, Serius, following Phaethusa to the banks of the Po river, Moffet uses the historical account of the introduction of sericulture to the West. For a detailed account of how silkworm *ova* were brought to Justinian by two monks, see G. F. Hudson, pp. 103–34.

70.13–24. Italy inherited the Byzantine silk industry, through Sicily; silk manufacture reached Spain as a result of the Saracen conquest of Persian Syria. There is an account of the Florentine industry in Edgcumbe Staley, *The Guilds of Florence* (1906) 204–35.

72.16. *from brutish beasts*. In the *New Arcadia,* Pamela claims that "to [human persons] only is given the judgement to discerne Beautie" (*Works of Sir Philip Sidney*, ed. A. Feuillerat [1912] 1.403). Alternatively, Moffet may be referring to the Aristotelian definition of man as the laughing animal (*De partibus animalium,* 673a1).

74.9. *lover wise*. It was a commonplace that love, being blind, could not distinguish between Hecuba and Helen (or "Astarte," that is, Aphrodite). See J. B. Bamborough, *The Little World of Man* (1952) 141.

Glossary

Abbreviations

*   This passage is cited in the O.E.D.
+   This is the first citation in the O.E.D.
†   This antedates the first citation in the O.E.D.
ded.   Dedicatory verses

ARTIST artisan; 70.4

BELLIES-FARMER collector of excrement; 60.14 [not recorded in the O.E.D.]
BETTER grow better; 68.8†
BIAST decorated with strips of cloth cut obliquely to the woven texture; 49.15
BLAINE inflammatory swelling; 52.3
BLOE deposit eggs; 53.19
BOTTOM silkworm's cocoon; 8.21† etc.
BRIGANDINE coat of armor; 36.3

CANCKAR ulcer; 52.5
CARRY-CASTLE elephant; 34.12*
CATTEL, CATTLE (silkworms) kept for profit; 40.22 and 55.12
CHALLENGE claim; 24.19
CHARME entreat (a person) by some potent name; 16.14+
CHEEKE upright of a door; 12.10
CHEESILL chisel; 14.22
CHINCKT cracked; 11.3+
CHUFF morose person; 5.15
CHURRING low whirring sound; 73.13+

CLARICALL clavichord, a keyboard instrument; 73.24
COAST 1. keep alongside; 32.5; 2. sail near the coast; 34.10
COFFIN cone; 62.15
COLONY discrete group; 57.10
CONTROUL instruct; 42.13
COOKE utter note of a cuckoo; 50.20+
COUNTERPEISE offset; 74.21
CREEPER worm; 18.12
CROOKE (of a wall) take a crooked course; 10.5

DEDALIAN skilful, ingenious; 47.11
DEFILE defilement; 2.21 [this sense not recorded in the O.E.D.]
DELAY quench (thirst); 13.1
DISMALL calamitous; 37.9*
DISTAINTED uncorrupted; 44.8+ [the O.E.D. glosses incorrectly as 'infected']
DISTIL trickle; 73.5
DISWITTED crazed; 34.9+
DOOM personal judgement; 42.14
DOUNE feathery substance; 62.17† and 63.17
DROP (of eggs) be laid; 66.15

ELMING belonging to an elm tree; 56.7*
EMPAIRE suffer injury; 39.11
EPIDEMIAN epidemic; 56.19

FALL 1. (of animals) be born; 47.13; 2. (of eggs) be laid; 66.18
FANNE fan-shaped leaf; 3.10+
FARMAR, FARMER 1. one who practices agriculture; t.p., *ded.* 16+, and 50.9; 2. one who purchases the right to collect; 60.14
FILE to defile; 20.4
FLAGGE long slender leaf of a plant; 34.19*
FLIE 1. any winged insect; t.p.* etc.; 2. speed-regulating device in clockwork machinery; 35.18+
FUME to perfume; 69.2

GLUE spittle; 11.6 and 65.21
GOOG'D gouged; 14.22*
GROVEL grow close to the ground; 51.20†
GUARD ornamental trimming; 23.19

GLOSSARY

HACKEL flax-comb used for separating fibre from the woody core of the stalk; 4.15*
HAGGESS magpie; 44.3+
HERSE coffin; 19.2†
HOLLA cease!; 37.1
HURRING buzzing noise of an insect; 73.13*
HURTELL swelling on the skin; 74.13

INDUSTRY ingenuity; 20.17
INURLD adorned with a border; 49.16+

KAFER chafer, a destructive beetle; 53.17†
KENNING range of sight; 15.12*
KRIME cold wind; 56.5+

LEVANT STUFF coarse, heavy silk; 63.18
LINGUIST noisy bird; 43.12*
LOOKER eye; 16.19 [not recorded in the O.E.D.]
LURE device for recalling hawks; AT LURE under control; 52.5*

MATE prove superior to; 42.14
MATELESSE without a partner; unwed; unparalleled (all three senses apply); 13.17
MEAGERNESSE leanness, emaciation; 55.23+
MEED heal; 31.7+
MENSTRUOUS menstrual; 67.3*
MOARE root; 6.4*
MOULDRESSE female creator; 47.11+
MULTIPLIER alchemist; 45.9
MURRY purple-red; 28.11

NEIGHBOURSHIP state of being a neighbor; 9.18*

ORIS fleur-de-lis; 34.19 (sidenote)
OVALL egg-shaped; 18.15*
OVERT open ground; IN OVERT in the open; 50.21+
OVER-WHEELE overturn; 59.15+

PAPRED covered with paper; 56.10+
PEACHY having a delicate pink flush; 28.7+

PURLD embroidered with gold or silver thread; 49.15
PYRAMIDALL pyramid-shaped; 45.9 (sidenote)+

QUARTER OUT divide into parts; 55.21*
QUILL hollow stem for winding yarn; 62.22

RAMP bound; 12.19†
RAYE moral influence; 37.19†
RECORD remember; 45.17 and 65.7*
RECREMENT superfluous portion; 26.10+
RETROGRADATE (of planets) go backward in the zodiac, so appearing to travel from east to west; *ded.* 8
REVOKE forfeit; 22.6 [this sense not recorded in the *O.E.D.*]
ROUND surround; 60.23*
ROWELL small wheel or pinion; 35.18+

SAGE presage; 42.12 [this usage not recorded in the *O.E.D.*]
SCANNE sum up; 25.5 [this sense not recorded in the *O.E.D.*]
SCORN prove superior to; 41.18†
SCOURE evacuate the bowels; 60.9 (sidenote)
SECUNDINE inner coat of a cocoon; 64.13+
SHADOW 1. to shade (as in drawing); 62.6†; 2. projecting shade of a woman's headdress; 68.20
SHINE sunshine; 13.19†
SHROWDE branches of a tree; 55.7
SITTER bird sitting on eggs; 24.22†
SKARRE drive away; 59.18
SKIPPER locust; 25.3*
SOULE relish, sowl; 54.23*
SPINSTER spinning silkworm; 27.16†
SPITTING sprinkle of rain; 55.11†
SPUNNE spun yarn; 75.6†
SQUIRE set-square; 35.5*
STONES kidney- or gall-stones; 52.12
STOOP (of a hawk) descend to the LURE (q.v.); 52.6
STUDY be in doubt or perplexity; 65.9

TIMBER-WORM worm living in cut timber; 23.9*
TINCTURE (in alchemy) spiritual principle infused into material things; 68.2+

GLOSSARY

TINSEL gauze made of cloth embroidered with gold or silver thread; 68.19
TITLING hedge-sparrow; 32.4
TOUSE tease (with a flax-comb); 4.16+
TRAVAILE passage (of time); 8.19
TREAD (of the male bird) copulate (with); 24.8 and 38.3*
TROUCHMAN interpreter; 10.2
TWISTER one who twists together the ends of the yarn of the new warp to those of that already woven; 69.18*
TYRY-TIRY-LEERER lark; 50.19+

UGH yew; 67.18
UNDESIRED not wished for; 38.15+
UNRIFE unusual; 4.14 [not recorded in the O.E.D.]
UNWHET cause (something) to lose its sharpness; 55.19+

WATCHET sky-blue; 28.4
WATER-WAVED (of textiles) having a wavy lustrous damask-like pattern or finish, produced by sprinkling with water and passing through a calendar; 3.14+
WEETE know; 53.13*
WHEELD having wheels; 30.4†
WHITED (of an egg) having albumen; 66.20+
WIHY whinny; 7.15
WINDER one whose occupation is to wind yarn; 69.17*
WINDLE wind; 1.8*
WITTOL half-wit; 34.16*
WOLFE malignant disease, usually on the leg; 52.5
WOODY inhabiting a wood; 14.2*

YOLKED having a yolk; 66.20*

Index to the Commentary

Aelian 36.15
Agrippa, Cornelius (of Nettesheim) 45.2
Alanus de Insulis 47.9–14
Ammianus Marcellinus 19.10
Anacreon (pseudo-) 36.21–22
Andrewes, Lancelot 34.9–37.24
Antiochus IV Epiphanes 22.11
Aristotle 40.9, 72.16
Ausonius 45.17–20

Bacon, Sir Francis 20.9, 38.18
Bamborough, J. B. 74.9
Bonfinis, Antonius de 68.9–16
Browne, Sir Thomas 20.9, 34.9–37.24, 39.1, 43.18, 61.17–24

Caius, John 56.19
Camerarius, Joachim, Jr. 61.17–24
Cicero 12.7, 46.17
Coccius, M. A. (Sabellicus) 11.19
Columella 51.5
Conti, Natale (Natalis Comes) 8.6, 18.3, 46.14, 64.18
Cooper, Thomas 16.7

Dalington, Robert 64.10
Daniel, Samuel ded. 9–11
Davies, John, of Hereford 1.2–3
Dee, John 35.15, 47.1–3
Diogenes Laertius 22.13, 63.21–24
Dodoens, Rembert 6.3–4

Du Bartas, Saluste 35.17, 48.13, 49.1-8, 50.19, 52.18
Duncan, E. H. 44.10
Dyer, Sir Edward 37.9-16

Fernelius, Joannes (of Amiens) 69.10
Fisher, A. S. T. 35.15
Forbes, W. T. M. 40.9
Fraunce, Abraham 40.17-21

Gentili, Scipio 1.2-3

Herbert, Mary (Countess of Pembroke) ded. 2-4, 1.2-3, 40.17-21
Herodotus 36.15
Hesiod 46.14
Homer ded. 9-11, 11.9, 29.8
Hooper, W. 62.21
Hudson, G. F. 19.10, 70.5-8, 70.9-10
Hutton, James 36.21-22
Hyginus 5.19

Jonson, Ben 44.9, 44.10, 45.9, 45.17-20, 46.20, 67.21

Lampridius 70.5-8
Lannoy, Cornelius 47.8
Lazarellus, Ludovicus 64.8
Leo, Joannes (Africanus) 52.19
Libavius, Andreas 46.1-6, 47.1-3
Lucretius 47.9-14

Macrobius 8.6, 24.5, 30.21-22
Maier, Michael 44.1, 67.21
Mantuan (Battista Spagnoli) 37.1
Milton, John 12.22
Monardes, Nicolaus 52.18

Orta, Garcia ab 52.18
Ovid 5.19, 9.9, 11.24, 14.3, 48.2, 58.10

Paracelsus 45.17-20, 46.1-6
Pausanias 68.17-24

Pererius, Bendictus 29.21
Peterson, Roloff 47.8
Pindar 31.13
Platt, Sir Hugh 9.19
Pliny the Elder 3.11, 8.17, 20.9, 25.1, 35.2–3, 36.17, 39.1, 41.17–44.8, 42.7, 51.4, 51.9–16, 51.17–24, 52.1–8, 54.19, 68.9–16
Plutarch 3.22, 24.5, 36.15, 43.15, 54.9–16
Propertius 20.11
Puteanus, Erycius 25.7
Puttenham, George 4.13

Ramus, Peter 35.17
Regiomontanus (Johann Muller) 35.17
Richerius, L. Coelius (Secundus) 23.9

Scoular, Kitty 34.9–37.24, 37.9–16, 51.1–52.24
Servius 45.17–20
Shakespeare, William 65.20
Sidney, Elizabeth 1.2–3
Sidney, Sir Philip ded. 2–4, ded. 5, 11.22, 72.16
Sidney, Robert 1.2–3, 45.9
Spenser, Edmund 1.2–3
Staley, Edgcumbe 70.13–24
Statius ded. 9–11
Synesius of Cyrene 18.15–20

Tillyard, E. M. W. 25.7
Topsell, Edward 43.18
Tubbe, Henry 69.10
Turba Philosophorum 47.1–3
Tzetzes, Joannes 39.6

Vaughan, Henry 61.17–24
Vergil ded. 9–11, 45.17–20, 46.22
Vergil, Polydore 70.9–10
Vida, Marcus Hieronimus 5.1, 5.10, 47.4, 48.1, 64.8, 70.9–10
Vopiscus 68.15

Welti, Manfred E. 38.10
Williams, A. L. 3.11

The Renaissance English Text Society
Council

PRESIDENT, *Arthur F. Kinney,* University of Massachusetts, Amherst
VICE-PRESIDENT, *A. S. G. Edwards,* University of Victoria
SECRETARY, *Suzanne Gossett,* Loyola University of Chicago

Thomas Berger, St. Lawrence University
A. R. Braunmuller, University of California, Los Angeles
Mario A. Di Cesare, SUNY-Binghamton
David Freeman, Memorial University of Newfoundland
Richard Harrier, New York University
Clarence Miller, St. Louis University
Janel Mueller, University of Chicago
Richard J. Schoeck, University of Colorado

International Advisory Council

Dominic Baker-Smith, University of Amsterdam
Richard S. M. Hirsch, Cambridge, England
K. J. Höltgen, University of Erlangen-Nürnberg
M. T. Jones-Davies, University of Paris-Sorbonne
Sergio Rossi, University of Milan
Germaine Warkentin, Center for Reformation and Renaissance Studies, Victoria College, University of Toronto
Henry Woudhuysen, University College, University of London

EDITORIAL COMMITTEE for *The Silkewormes and their Flies*
  Thomas Berger
  A. R. Braunmuller
  Arthur F. Kinney, Chair

The Renaissance English Text Society was founded to publish scarce literary texts, chiefly nondramatic, of the period 1475–1660. Originally during each subscription period two single volumes, or one double volume, were distributed to members. Beginning in 1978, with the publication of Series IV, members are billed $15 annual dues regardless of whether there is a volume published during the year; all subscriptions are used for printing and publishing costs, and members will be credited with the amount they have paid toward each series when it appears.

Subscriptions should be sent to Arthur F. Kinney, Department of English, University of Massachusetts, Amherst, Mass. 01002, USA. Institutional members are requested to provide, at the time of enrollment, any order numbers or other information required for their billing records; the Society cannot provide multiple invoices or other complex forms for their needs.

Copies of past publications still in print may be purchased from Associated University Presses, 440 Forsgate Drive, Cranbury, New Jersey 08512, USA. Current and future volumes will be available from MRTS – LNG-99, SUNY, Binghamton, NY 13901.

FIRST SERIES

Vol. I. *Merie Tales of the Mad Men of Gotam* by A. B., edited by Stanley J. Kahrl, and *The History of Tom Thumbe*, by R. I., edited by Curt F. Buhler, 1965.

Vol. II. Thomas Watson's Latin *Amyntas*, edited by Walter F. Staton, Jr., and Abraham Fraunce's translation *The Lamentations of Amyntas*, edited by Franklin M. Dickey, 1967.

SECOND SERIES

Vol. III. *The dyaloge Called Funus*, a translation of Erasmus's colloquy (1534), and *A Very Pleasant and Fruitful Diologue called The Epicure*, Gerard's translation of Erasmus's colloquy (1545), edited by Robert R. Allen, 1969.

Vol. IV. *Leicester's Ghost* by Thomas Rogers, edited by Franklin B. Williams, Jr., 1972.

THIRD SERIES

Vols. V-VI. *A Collection of Emblemes, Ancient and Moderne,* by George Wither, with introduction by Rosemary Freeman and bibliographical notes by Charles S. Hensley, 1975.

FOURTH SERIES

Vols. VII-VIII. *Tom a Lincolne* by R. I., edited by Richard S. M. Hirsch, 1978.

FIFTH SERIES

Vol. IX. *Metrical Visions* by George Cavendish, edited by A. S. G. Edwards, 1980.

SIXTH SERIES

Vol. X. *Two Early Renaissance Bird Poems,* edited by Malcolm Andrew, 1984.

Vol. XI. *Argalus and Parthenia* by Francis Quarles, edited by David Freeman, 1986.

Vol XII. Cicero's *De Officiis,* trans. Nicholas Grimald, edited by Gerald O'Gorman, 1987.

Vol. XIII. *The Silkewormes and their Flies* by Thomas Moffet (1599), edited with introduction and commentary by Victor Houliston, 1988.

The Silkewormes and their Flies is here available for the first time in a modern critical edition, presenting in handsome facsimile the first georgic on the methods and rewards of sericulture. Written by the distinguished Elizabethan physician and entomologist Thomas Moffet, it was first published in an attempt to promote the planting of mulberry trees and the rearing of silkworms in England, in order to meet the demand for silk production.

Houliston's introduction discusses the poem's significance as the first Vergilian georgic in English, its connections with contemporary nature poetry, and its debts to the tradition of the mock encomium. He traces England's interest in sericulture, and provides a full bibliographical description of the original manuscript. His notes offer extensive commentary on sources, parallels, scientific and classical references. Scholars will be grateful for the glossary of unusual terms and early recorded usages, and the index.

Victor Houliston is currently a Lecturer in the Department of English at the University of Witwatersrand, Johannesburg.

mrts

medieval & renaissance texts & studies
is the publishing program of the
Center for Medieval and Early Renaissance Studies
at the State University of New York at Binghamton.

mrts emphasizes books that are needed —
texts, translations, and major research tools.

mrts aims to publish the highest quality scholarship
in attractive and durable format at modest cost.